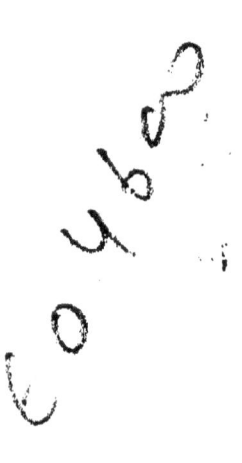

PRACTICAL ORCHARDING ON ROUGH LANDS

By Shepard Wells Moore

The New Werner Company,
Akron, Ohio.
1911.

Practical Orcharding On Rough Lands

"Trees are our teachers,
If we but read their silent lives aright,
Rooted in clay, they lift their heads toward the light."

"Teach me, Father, how to be,
Kind and patient as a tree."
—*Edwin Markham.*

Shepard Wells Moore, Gallipolis Ferry, West Virginia

Copyright 1911
By Shepard Wells Moore

TO ALL TREE-LOVERS.

"Who loves a tree, he loves the life that springs in star and clod;
He loves the love that gilds the clouds and greens the April sod;
He loves the wide Beneficence. His soul takes hold on God.

A tree is one of Nature's words, a word of peace to man,
A word that tells of central strength from whence all things began,
A word to preach tranquillity to all our restless clan.

Ah! bare must be the shadeless ways, and bleak the path must be,
Of him who, having open eyes, has never learned to see,
And so has never learned to love the beauty of a tree.

'Tis well for man to mix with men, to drive his stubborn quest
In harbored cities where the ships came from the East and West,
To fare forth where the tumult roars, and scorn the name of rest.

'Tis well the current of his life should toward the deeps be whirled,
And feel the dash of alien waves along its channel swirled,
And the conflux of the eddies of the mighty-flowing world.

But he is wise who, 'mid what noise his winding way may be,
Still keeps a heart that holds a nook of calm serenity,
And an inviolate virgin soul that still can love a tree.

Who loves a tree, he loves the life that springs in star and clod;
He loves the love that gilds the clouds and greens the April sod;
He loves the wide Beneficence. His soul takes hold on God."
—*Sam Walter Foss,* in *"Songs of War and Peace."*

Apple Tree From Seed Planted in 1790 by Mrs. Geoffrey Haynes Now Measures 10 Feet in Circumference. Top Spreads Over 50 Feet and is 40 Feet High. Near Sinks' Grove, Greenbrier County, West Virginia

THE APPLE TREE.

"Great blessings on the man who planted thee,
    O apple tree,
Within my garden's safe enclosure kept,
    While winter slept,
Then went thou summering with birds and bees
    And other trees.
First came thy springtime joy, when rills
    Sang to the hills,
And thou didst laugh and shake thy bells of bloom,
    And made sweet room
For wandering breezes, murmuring some love song
    The whole day long.

Great blessings on the man who planted thee,
    O apple tree,
For when the winter storm wails round my room,
    Thy springtime bloom
Lives in the rosy apples on my table here;
    Their fragrant cheer
Gives gentle hint of how they grew, and grew,
    When sun and dew,
And shadows falling in the evening gray,
    Made up the day.

Great blessings on the man who planted thee,
    O apple tree,
With what sweet thought he hoped that thou wouldst grow;
    And so and so
The slender sapling climbed a little way
    By night and day.
All silently and hidden, tender root
    Supplied the shoot;
The long, lithe limbs reached further, further yet,
    With green buds set,
And lo, a great tree, on whose leafy crown
    The sun shone down,
Till every apple, green and garlanded,
    Blushed rosy-red.

Great blessings on the man who planted thee,
        O apple tree,
I love the man who builds not for himself, alone,
        Some little home;
Who plants a tree knows not, nor thinks for whom
        That tree may bloom;
But some one coming after him, will bless
        His thoughtfulness."

TABLE OF CONTENTS

CHAPTER I.
The First Essentials for Success...................... 19
Every man cannot become a successful orchardist—Love for money alone is not enough to ensure success—Never a more serious time—Must have a love for our work in order to get the most out of it, either pleasure or profit.

CHAPTER II.
Orcharding as a Business......................... 25
Recognized as a profession today—There is a vast difference in growing Fruit for Home Use and in growing it for the great markets of the world—It seems to be the desire of every man to grow, handle and eat fine fruit—This is a day of specialties—Preparation differs in different sections—Large companies formed for the planting—Not restricted to any particular section—The Nurserymen important factor—Will production be overdone?—A good time to enter the Business—The East a good place—Study location.

CHAPTER III.
Location 33
Choosing a location—The chief factor in determining—Is climate—Apple zone divided—Allegheny Section—Lake Ontario Section—The would-be planter—The choice of location—Transportation—Water supply—Site—Choose the soil—Altitude—Prolong the seasons—Advantage of the mountain sections over the more level.

CHAPTER IV.
Drainage 47
Should receive more careful attention—Difference between moist and water soaked soils—Pays to drain lands—For crops—For orchards—Drainage necessary on steep lands—Root Rot—Air or frost drainage—The comparative height—Not altitude—Difference in frost and freezes—The successful fruit grower of the future.

CHAPTER V.
The Aspect 57
A very great difference of opinion—Experience and observation—Various slopes—Strength of soil—Moisture holding capacity—Unproductive Soils—Availability—Plant food—The warmth of the soil—Protection from wind.

CHAPTER VI.

Windbreaks 65
Difference of opinion—Much emphasis—Arguments in favor of—Objections—Belts of timber—May shelter—Sometimes interferes with growth—Set and build trees so they will protect—Prune for protection against wind.

CHAPTER VII.

Preparation of the Site 73
Should have thought—Should be a pleasure—Different methods—Different kinds of land—Object, root to occupy entire surface—Prepare in fall—Dig holes early—Watch drainage.

CHAPTER VIII.

Laying Off the Orchard 83
The plan of setting—Difference in the number of trees per acre—The operation of laying off the orchard—Distance apart—Strength of the land—Habits of growth—Fillers—Arrangement of varieties.

CHAPTER IX.

Selection and Care of the Nursery Stock............ 97
Where to purchase—Visit nursery—Poor nursery stock—A good tree—Well grown according to varieties—Free from insects—Aphis knots—Fungous troubles—Manner of propagation—Grimes Golden—What aged tree to plant?—Comparative hardiness in fall and spring setting—Too early delivery—Stripping off the leaves—Purpose of the leaf—Treatment of dried and shriveled trees—Heeling in—Labelling—Mice trouble—To prevent growth.

CHAPTER X.

Planting the Tree 115
When to plant?—Soil in good condition—When trees are moved from the nursery—Arrangement of varieties in the orchard—Prepare tree for planting—Depth of planting—Puddle before setting—Position to set tree—Sunscald—Longest root as anchor—Heavy side of tree—Dust mulch around the tree.

CHAPTER XI.

Care and Cultivation 133
  Why we cultivate—Cultivate to set free plant food—To deepen a soil—Cultivate to increase moisture holding capacity—To retain moisture—Cultivate to hasten decomposition of plants—Cultivate to destroy plants—To develop plants—Growing plants—Growing plants to supply humus—Growing plants to control the growth of other plants—Terminal bud—Different methods necessary in different sections—Care of the orchard—Crops first year—Care of trees first season—Kind of wrappers to use—Winter care of the orchard—Crops to grow in young orchards—Summer cover crops—Winter cover crops—Treatment of cover crops—Mulches—Dust mulch—Foreign mulches—Mulch grown in orchard—A growing mulch—Sod mulch—Self-mulched trees—Care of trees.

CHAPTER XII.

Pruning 189
  The planters ideal—The ideal tree—An ideal fruit tree—Low-headed trees—Habit of growth of tree—Purpose of pruning—The first effect—Pruning to modify vigor—Checking growth to cause fruitfulness—Pruning to produce larger and better fruit—Pruning to remove unnecessary parts—Pruning to remove injured parts—Pruning to renew bearing wood—Pruning to renew bearing wood in old trees—Treatment of water sprouts—Pruning only for form and size—Pruning to form the head of the tree—Pruning to remove insect infested parts—Pruning to bring into manageable shape—Pruning with regard to the location and formation of the flower buds—Art and science of pruning as related to flower buds—When to prune—How to prune—Where to cut—What tools to use—Cutting large limbs—Treatment of wounds—Remove the brush—Summary.

CHAPTER XIII.

Spraying 247
  How and when to spray—Protection—Thoroughness—What shall we spray with—Insects—Spraying for scale—Appliances—Pump—Length of hose—Nozzle.

CHAPTER XIV.

Picking, Packing and Marketing 266
  When to pick—By what shall we be governed in picking—Make several pickings—How to pick—What to use in picking—Packages—Marketing—When to market.

Practical Orcharding On Rough Lands

OHIO RIVER HILLS
Courtesy J. F. Cunningham

"And soon or late to all that sow
The time of harvest shall be given;
The flowers shall bloom, the fruit shall grow,
If not on earth at least in heaven."

INTRODUCTION.

This is an age of advancement. As a people we are proud of our achievements. With eager watchful eye we note the ever varying customs of those about us, even the fashions that come and go with each recurring season, as so many fast flying shuttles. Each shuttle carries its own thread of special hue and cannot fail to leave its trace in both warp and woof of the nation's fabric. Yes, we are quick to notice all these things, but slow to realize the changes that have come about in our country, in our soil, in our crops, in our markets and their demands.

We read with interest of the discovery of our country, and follow excitedly the explorer as he searches out the wonders of a new land. With the prospector, we seem to be climbing the mountains; tramping through the trackless forests, and fording the swollen streams. We picture with great vividness his day dreams of the future as he viewed the country from some lofty height, and in his imagination, located a city at the junction of these streams, and another at yonder gap in the great mountain range. We seem to see in our mind's eye, as he did, the smoke curling up from the foundries, and we think we can hear (as he did in his imagination) the buzz of wheels in mill and factory. We are apt to be envious of those men whom we think of now as speculators, when we read of their dreams of great coal veins and immense railroad systems carrying to the markets nature's stores of wealth.

This is only a backward glance at the development of our country, and while it may be a good thing to look back occasionally, let it be only that we may have a clearer vision of the future.

It shall be the purpose of this book to try to

show the young man of to-day that all the development is not completed. The field is as ripe for the harvest now as in the past, but the field is to be occupied by another class of workers; men who are willing, by their energy and industry, to transform our brush covered hills into profitable orchards. These men are making history just as truly as did those of days gone by. They also have had day dreams of lands and opportunities that have been passed by or overlooked Yes, in the rush for the wealth of our natural resources, the surface of much of our rough and rolling lands has been neglected, and now there appear bright dreams of happy homes, nestling among the orchard covered hills and mountains, which are pink with peach, and white with the apple blossom, and all the breezes (as they bring to us happy voices of childish glee) seem laden with sweet perfume.

The busy knock-knock of the cooper is heard from early morning until late at night, as he drives down the hoops on the thousands of barrels that are to carry the product of the orchards to the great markets of the world.

Railroads are breaking in on quiet valleys. The shrill whistle and the clouds of smoke tell us they are hauling away the timber from the mountain sides, giving place for thousands and thousands of fruit trees. This is not a mere dream. We see acres and acres of rough land now in orchards, which are yielding handsome returns for the time and money expended in their development. The people in general seem to realize that we are approaching the time of more intensive husbandry, and are beginning to appreciate, as never before, the value of our rough lands for orchard purposes. The hundreds of young men who come with their questions as to location, site, preparation, selection of trees, choice of varieties, planting, care, cultivation, pruning, spraying, picking, packing and marketing; represent a great army of fruit growers of the future.

Their constant tramp, tramp, as they march against the many difficulties which present themselves has caused the writer to take up this work and attempt to encourage, and possibly to aid them, by a few suggestions.

These pages have been written after years of experience in orcharding on rough lands, and while it is not our purpose to advise any one to hunt out waste land for orcharding, we will try to set forth some of the possibilities and even advantages of such rough or rolling lands as may be at the command of the reader.

If we shall cause some one to plant and care for only a small home orchard, which would give his family a bountiful supply of fruit, (which every family should have) and thereby create in the boys and girls a love for good fruit and a desire to plant and care for trees, it may be a means of advancing horticulture, as their work and its effect may be passed on and on to future generations.

Should we be able, by giving to the public our experience, to keep even one person, who like the writer, found it necessary to utilize rough lands for orcharding, from making some of the many mistakes which crowd in upon us as fruit growers, we shall feel that our efforts have not been in vain, as has been so well expressed in the following lines:

"I may not reach the heights sublime;
My place is lowly and unknown.
But if I've caused the light to shine
Across some pathway, dark and lone;
If some one called me kind, and then,
Another found in me a friend;
If but one wanderer on Life's way,
Would pause beside my grave, and say,
'He did his best the world to make
A sweeter place, for mankind's sake,'
Or some bent form its step would stay
To whisper, 'He helped me on my way';
Or one in sore distress or need,
Remembering me, would for me plead;
I shall not miss the laurel crown

That victors wear, nor cap nor gown
The great achieve, for service done.
IF I HAVE SERVED—I shall not moan
That I have not been better known."

—*Shepard Wells Moore.*

ORCHARDS ON ROLLING LANDS
Courtesy J. H. Hutchinson

CHAPTER I.

THE FIRST ESSENTIALS FOR SUCCESS.

"The man who wins is an average man;
Not built on any peculiar plan,
Not blessed with any peculiar luck;
Just steady and earnest and full of pluck."
—*Charles R. Bartlett.*

In order to succeed we must have a love for our work. This is necessary in any business and especially is it true in orcharding, for orcharding is a business which requires years of enthusiastic work to attain profitable results. The high prices of certain crops, such as corn, wheat or tobacco, may induce some to engage in their production, and although they have no special love for the work they may be able to keep up their enthusiasm for the short time necessary to realize a profit from their efforts, and then change their line of work. But how different with the orchard business. The high price of fruit may encourage some to plant trees by the thousands, and then if interest be lost by a series of years of low prices for the product, or from ravages of insect pests, early or late frosts, or from any cause, climatic or otherwise, all may be lost. Yes, and the business discredited. So if there is not in the bosom of the planter a love for the work, the business is likely to suffer sooner or later.

Not every man can become a successful orchardist any more than he can become a successful stockman, lawyer or physician. So we should study our likes and dislikes. If we find we love the busy streets of a crowded city rather than the quiet field, then work in the city. If the windows with their displays of bright colors are more attractive to us than the trees with their delicate wreaths of blossoms, then be a merchant. If the jewelry and clocks in the show cases have more attraction for us than those lovely colored and beautifully finished cases, which we call fruits, then listen to the tick-tick of the clocks. If the building of houses is more attractive than the building of plants, and especially trees, then use your saw on dry boards rather than on living trees.

If the hum of the factory and machine shop has greater charm than the song of the birds, go there. If we like the rush and the roar of the railroad train rather than the sound of the spray pump and engine, there is work waiting for such.

In short, if we do not love to see the swelling buds, and if the unfolding leaves of our trees do not interest us in their quiet but important task, from early spring until late autumn, when they take on their gorgeous colors, thereby showing that their work is nearing completion, if we cannot see beauty even in the falling leaves, if we do not love nature and her wonders as they are daily shown in things about us, but simply attempt to grow fruit for the dollar and the dollar only, we may succeed; but we will lose much pleasure and encouragement that should be ours, for it is the love for one's work that gives birth to hope within his breast—and hope is as a bright star that beckons us on and on until we reach the goal

> "Behind the cloud the starlight lurks,
> Through showers the sunbeams fall,
> For God, who loveth all his works,
> Has left his hope with all."

Spring Time in the Orchard
Courtesy Leo Jellinek

THE APPLE ORCHARD.

"Of the great beauties of the farm—
The one that has the foremost charm
The apple orchard leads them all,
From early spring to latest fall.
The budding trees of pink and white,
The whole world shows no fairer sight.
An apple tree's full bloom will stand
Beyond all rivals, broad and grand.

The honey bees, rejoicing find
Its blooming sweetness to their mind,
They come from far, they come from near—
The early harvest of the year.
At length the baby apples show
Amid the green leaves, growing slow;
A promise of a tempting treat,
Which rain and sunshine kindly greet.

Those red-striped apples, tempting, rare;
No golden orange quite so fair
Though tropic trees have long been mine—
For apple orchards still I pine.
That orchard to my childhood known,
No tropic fruits can half atone.
The early bloom—the red-cheeked fruit,
Are visions which my dreams salute."
—*Coleman's Rural World.*

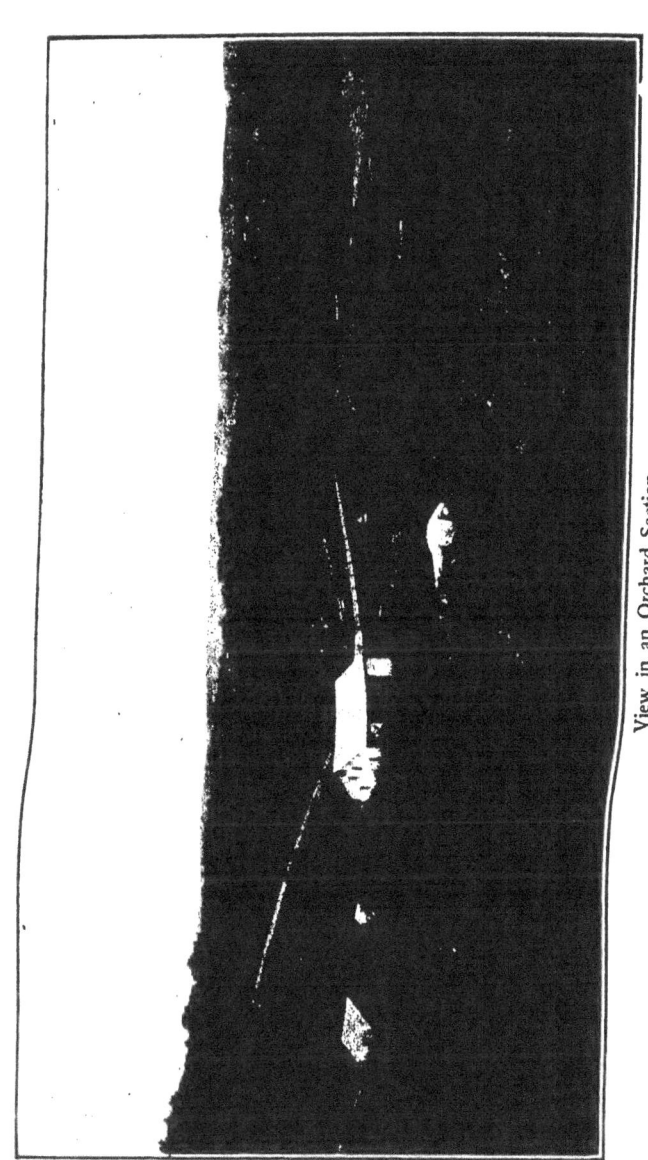
View in an Orchard Section
Courtesy S. W. Fletcher

ORCHARDING ON THE MOUNTAIN
Courtesy Leo Jellinek

CHAPTER II.

ORCHARDING AS A BUSINESS.

"It is not just as we take it,
This mystical world of 'ours,
Life's field will yield as we make it,
A harvest of thorns or flowers."
—*Alice Cary.*

There is never a more serious time in one's life than when he attempts to decide upon his life work. That this decision or choice should be carefully studied from all view points will be accepted without argument. While this is true in the choice of what are commonly known as professions, namely law, medicine, etc., it is none the less true when we come to choose among the branches of agriculture. Whether we are to grow grain, breed and feed stock, or to spend our time in the study of the soil and the various plants that it will grow. These are all recognized as professions today just as much as law or medicine. In order to succeed in any one of them we must apply business methods just as surely as must the banker or merchant. While this is true of all the various lines of agriculture we believe it is particularly so of ORCHARDING AS A BUSINESS.

There is a vast difference in growing fruit for home use and attempting to produce it to supply the great markets of the world. That there is a fascination in the production of fruit with which to provide one's own table none will deny. This pleasure seems all the greater if we have planted and cultivated the trees with our own hands in our own garden or orchard. As has been so well said

" 'Tis strange how we learn to love the things we plant and tend,
Every tree in that whole orchard seems like some dear old friend."

It seems to be the desire of every man—whether he lives in the country or city—to grow, handle and eat fine fruit. We can scarcely enter a bank or store or ride on a railroad train but what we hear the men engaged in these various lines of work express a desire to get out on the farm and grow fruit. We find the merchant, the lawyer, the banker and the man in the shop who by years of hard work and rigid economy have laid by a few thousand dollars looking forward with great pleasure to the day when they may either have an orchard all their own, or at least hold some stock in one.

This is a day of specialties, and while we find men of all classes and professions interested to a certain extent in the production of fruits they—many of them—would fall far short of making a success of orcharding as a business. While this interest in rural life, especially in fruit growing has existed for a long time it is even more noticeable today than ever before. City men who have made their millions in other pursuits are forming stock companies, buying large tracts of land and preparing them for planting fruit trees.

This preparation differs in character as to the section of country in which the land is located. In some cases it has been the arid lands of the west, where the large scattering stones have to be removed and the land levelled ready for irrigation. In many cases a great ditch must be constructed in order that the water may be brought from some stream which is fed by the melting snow on top of the mountains. Again we find these companies taking rough and rolling lands after the most of the forests have been cut away by the lumbermen and clearing them by the hundreds of acres and planting great orchards. Such plantations have increased greatly in the last few years, both in size and in numbers. This increase has not been restricted to any particular section, but is very noticeable all over the fruit belts from north to south.

So general has been this increase in planting that large nursery companies have been formed in the various sections of the country in the attempt to suppy the constantly increasing demand for fruit trees. The nurserymen and their work are very important factors in orcharding as a business.

The question is often asked, and very naturally too, if we do not believe that fruit production will be over done? And we are ever just as ready to answer "no." There will always be room at the top, or sale for first class fruit at fair prices. We should not fear over-production, but let us rather guard against over-planting. Neglect or carelessness in the methods employed may cause the markets to be filled with fruit of such poor quality that it may find slow sale at such low prices that will not prove profitable to the grower.

A GOOD TIME TO ENTER THE BUSINESS.

When we consider orcharding as a business there never has been a time that it seemed more promising than the present, and surely there is no country or section that offers to the young, wide-awake business man better inducements to enter the field than our own United States. Especially the eastern section, where land is cheap as compared with other fruit sections and yet so productive of fruits of good size, fine color and splendid quality. This is a section which has been considered by many as almost if not quite worthless—our rough and rolling lands—until in recent years when the success attained by some persistent, hard working men who have made orcharding a business has attracted the attention of the on-lookers. So we find men from all ranks and stations in life entering the field of fruit growing, some coming with hopes of great gain and little labor, others are seeking a safe and profitable investment, taking for their basis some phenomenal, or at least more than average crop, often not stopping to consider the time, money, or above all, the experience necessary to bring about such results.

Others are entering the business with a view of getting rich quick and we find them borrowing money on prospects that are years in the future. Then we have those whose health demand the open air and they are attracted by the beautiful sweet scented blossoms, the green leafy bowers and the red and golden fruits, all of which they think point to a good bank account, so they leave the office and the school room and embark in orcharding as a business. Frequently taking up their work as though they thought the only thing necessary to success was to plant the trees and the rest would naturally follow. Then we find men engaging in fruit growing because they love nature and want to come in close touch with her, by living among and watching the growing trees. They seem to get pleasure out of all their labor, from the planting of the trees to the gathering of the fruit. They watch with interest every change from the bursting of the buds to the falling of the leaves.

As we pass through the country we find a great many people engaging in orcharding in one way or another with almost as many and varied results as there are individuals. It is only when business methods have been applied that a high degree of success has been obtained. Orcharding as a business must be studied in all its phases, and when it is we should have no fears of failure. It is then that the most pleasure and profit can be realized.

We should study our location, soil and site as well as our means of transportation, then choose our varieties accordingly. We should give careful attention to the choice of the nursery stock, to the planting, care and cultivation of the trees as well as to the building of our trees, for we should think of pruning as the building of the orchard, and upon this depends our success. Our trees should have constant protection from insects and fungous troubles by thorough applications of the latest and best tried remedies given by our Experiment Stations and practical orchardists.

If we are to make orcharding a successful business we must not stop here, but we shall have to study the problems of picking, packing and marketing. Realizing that if we should fail in any of these, all that has gone before would be lost.

Orcharding as a business then means a careful study and constant watching of every detail, just as is necessary in any other business.

THE ORCHARD.

"I've planted trees, and these I call
 An orchard yet to be.
I wonder oft if e'er there'll fall
 Those apples ripe for me.

The trees are small, but growing fast,
 An orchard yet to be;
But as they grow the years fly past,
 And shorten life for me.

I picture to myself these trees,
 Grown large in coming years,
Their branches waving in the breeze,
 Red-coated fruit appears.

If they should ne'er bear fruit for me,
 I have two boys small,
And hope that they may live to see,
 Red-coated apples fall."

—*Frank Monroe Beverly.*

ORCHARDING ON WEST VIRGINIA HILLS
Courtesy J. H. Hutchinson

CHAPTER III.

LOCATION.

"How grand is the apple that grows by the gate,
We welcome the apple, be it early or late,
Yes, welcome the apples, its sweetness or tart
Outrivals all efforts of labor or art.
We may tire of oranges, bananas and grape
But never of apples, be they early or late."

By location is meant that particular part or section of the country in which the planting is to be made.

Orchardists, when choosing a location, should not lose sight entirely of the geography of fruit growing, as the industry thrives best in certain geographical sections. That is, the business is not capable of equal development in all parts of the country.

The chief factor in determining fruit production is climate. We recognize three great fruit zones, and while there are no plainly marked lines which determine their boundaries, they are marked in a way by the kinds of fruits that are found in them; for instance, the cocoanut, the orange and the apple.

We find the apple zone divided by Prof. Waugh into five great belts; these in turn are indicated by certain characteristic varieties. For example, the Mississippi Valley section, comprising Ohio, Indiana, Kentucky, Kansas, Arkansas, Missouri and Illinois has the Ben Davis as its characteristic variety; while the Jonathan, Grimes Golden, Willow Twig, York Imperial, Rome Beauty and many others succeed.

The Allegheny section comprising the slopes of the Allegheny mountains, Pennsylvania, West Virginia, western Virginia, eastern Tennessee and western North Carolina, has as its characteristic variety the York Imperial, while Winesap, Grimes Golden, Mammoth Black Twig, Ben Davis and others succeed admirably.

The Lake Ontario section comprises north western New York, adjacent parts of Ontario and south eastern Michigan. Here we find the Baldwin, Northern Spy, Greening and Roxbury Russett at their best.

In view of the fact that climatic conditions are a ruling factor in the production of the various fruits, and the different varieties of fruit, we should decide upon the varieties we wish to grow, and then choose the location, for it would be a mistake to attempt to grow an apple requiring a long season for maturity (such as Ben Davis) in a section suited to the Spy or Baldwin, while the Baldwin, if grown too far South, will ripen early and fill the place of a late fall rather than of a winter variety.

Climate should have the most careful consideration of the planter when choosing a location; remembering if he wishes to grow certain varieties he should choose a location where the conditions are congenial to the development of that variety.

Neither should the would-be planter lose sight of the fact that sections do not produce fruit equally well all over their entire area. For example, we find in certain fruit sections, localities where there is scarcely enough fruit grown for home consumption, while in the same section there may be localities in which there are large commercial orchards, often times some local conditions being largely accountable for the success of that particular section.

The choice of a location for any business should receive the most careful consideration, and especially that of orcharding. As an orchard cannot be moved from place to place, as could many other enterprises when we find a mistake has been made in regard to the surroundings, whether this mistake is on account of climatic or other conditions.

TRANSPORTATION.—This is one of the most important considerations when locating the orchard. It is very expensive to haul large crops of fruit a long distance, as well as being injurious to the fruit. A long haul over bad roads is often the cause of the fruit reaching the consumer or storage in bad condition. So we should consider carefully our means of transportation in connection with the study of our markets. If possible, we should be near a good home market, one that would take our "seconds" at a fair price, for we are sure to have some of the "seconds" no matter how careful we are in growing our crops. Then we want a home market that will be willing to appreciate and pay for fancy fruit as well. It is absolutely necessary that we have easy access to the great markets of the world. If two lines of transportation were available, it would make the location all the more desirable. Competition is the life of trade.

WATER SUPPLY.—The water supply for spraying purposes should be carefully planned for, as we have long since passed the day when the question is asked "will it pay to spray?" We have now arrived at a point in fruit growing where the absolute necessity of spraying as a protection of the foliage as well as the fruit is recognized, not only by the orchardist but also by the storage man, and the consumer. Hence, water has come to be recognized as one of the necessary factors in the production of fruit, as it is the means of application or carrier of insecticides and fungicides, as well as the carrier of food materials that are taken by the roots from the soil in the form of moisture.

The grower should not fail when selecting a location for his orchard to take into consideration how much the demand for water for spraying purposes will increase as the trees attain age. When young a few hundred gallons will spray quite a number, but after they have grown fifteen or twenty years and are loaded with fruit, they may require more frequent applications. This, together with their large tops and heavy foliage, will necessitate the use of many times the amount of water that was used early in the growth of the orchard. Sometimes this kind of an oversight proves very expensive in after years, as it may mean a complete change of the water system. The water supply is much more of a problem on rough, steep lands than on those that are more level, as it is not practical to haul it any great distance, especially up hill.

This subject of water supply has been brought home to us so forcibly in our own orchard management, that we wish, if possible, to emphasize the importance of a bountiful supply in the beginning. As has already been said, it is hard for the beginner to realize how the demand will increase with the growth of the trees, and we are apt to depend upon most any supply that may be at hand, not thinking of the future.

After drilling a well, erecting a wind-mill, and building reservoirs, which are located on the orchard hill, 2000 feet distant from the well, and have a capacity of 3000 gallons, for the spraying of 75 acres of orchard. Then after establishing mixing stations on the various hillsides, where they would be the most accessible, and laying two feet below the surface a system of galvanized pipe with brass fittings, we thought we had one of the most complete water systems for orcharding in the country.

But after a few years' experience, together with the growth of the trees, we found we had made several mistakes. First, the main pipe, which was one inch, and had to carry the water from the well in the low land to the reservoir,

150 feet above the well and 2000 feet away, was too small. There being too much friction to overcome, thereby increasing the tax on power and pump. Then, wherever a stop and waste was put in, they should have been from a quarter to a half inch larger than the line, as the openings were too small to allow a full flow through the pipe. This has been particularly annoying in the distributing pipes between the reservoir and mixing tanks, as the pressure in these lines is not sufficient to force a strong enough flow through the smaller openings to furnish water as fast as the spray pumps could use it, thereby often retarding the work. In former years this mistake was not noticeable, but with the growth of the trees and the necessary use of more spray rigs, the mistake becomes more and more apparent.

It is well when possible, to choose a location with springs on the same level, or better still if the springs are on somewhat higher ground, so that the water may be carried by gravity over the orchard to the various mixing places. This will prove to be one of the advantages that the steep lands possess over the level sections. If the springs are on lower ground, or if running streams are to be resorted to, the hydraulic ram may be used to lift the water to the storage tanks. Storage tanks or reservoirs will be necessary no matter where the supply comes from, for there should always be a quantity of water at hand ready for mixing. If the ram is to be used, it should not be overlooked that it lifts only a small percentage of the water afforded by the stream, the rest being lost in the working of the ram. Sometimes wells are drilled from which the water is pumped, either by windmill or gasoline engine and forced to the reservoirs or tanks. If the windmill is used it will be necessary to provide larger storage tanks than if the engine is the lifting power. The engine can be started at any time, but the windmill should be allowed to run, so as to insure an ample supply of water whenever the weather is suitable for spraying, as one cannot afford to have any delay at this all important season.

SITE.—The site is that particular piece of ground upon which the trees are to be planted. In thinking of it there are many things to be taken into consideration. First, should be adaptability. Does the soil suit the varieties to be grown? This is important, and its importance is recognized more at present than in former years. We realize the fact that certain varieties succeed on certain soils. How are we to know whether or not the soil will suit? This is a hard question, and when we begin to search for information, where better shall we turn for more striking lessons than nature gives us in the forests. We recognize white oak land or black walnut land when we see it, simply by seeing certain kinds of timber growing on this particular soil. So as we go through the country and observe this fact we come to recognize these soils as adapted to the growth of certain trees, shrubs and vines. Then a good plan would be to examine the old trees or orchards where there are trees of the varieties decided upon, and after considering their location, soil, altitude, etc., draw a comparison, (for it is by comparison that we should study all these things) with this soil, and the soil of the site to be planted, and then be governed accordingly. This will prove much safer than the haphazard plan of planting any varieties on any soil, regardless of their likes or dislikes. If the varieties are not just what are wanted, then take some member of the same family, whose color, size, etc., may suit better. For instance, if the Winesap is succeeding, but if you object to them on account of the fruit growing smaller as the trees attain age, so that they are too small to meet the requirements of your market; then do as the stockman who likes Short Horn cattle. The strain he has may not be his ideal

A Stony Orchard That Has Been Profitable
Courtesy W. E. Rumsey

type, but he does not cross with another breed. Instead, by continuous selection of the type wanted, he will ultimately build up a herd of the desired type. If the Winesap suits the soil but does not suit the orchardist, then he has the family of Winesaps to choose from, such as Black Twig, Kinnaird's Choice, Stayman's Winesap, etc.

If the Ben Davis suits the soil, but the name is distasteful to the grower, as it is to some, then choose Black Ben or Gano.

The successful fruit grower of the future must pay more attention to the selection of the soil, and its adaptability to certain varieties. We find whole orchards of Ben Davis being planted where Baldwins would succeed, and vice versa, as well as many other varieties that might be mentioned. In the future it will not be enough to simply know that pears grow best on clay, while peaches succeed best on sandy or gravelly soil, but we shall all come to recognize the adaptability of certain soils to certain varieties of apples, just as surely as we have for years, the adaptability of the various varieties of strawberries to certain soil. Recognition of this fact has been more general by the growers of strawberries than by the growers of the tree fruits, perhaps, on account of the frequency of the plantings. As well as because of the few years necessary to produce new generations of the same varieties. Or even entirely new varieties, from seed which may be the result of crossing some that did not exactly suit the soil upon which they grew, while the new one (or the seedling) may suit it exactly.

There is no doubt that each family of apples (so to speak) has its likes and dislikes. This opens to the young man of the present and future generation a vast field of investigation— a field rich with opportunities for the careful, studious, observant orchardist, in which he may do a great work, not only for the orchardists of the present, but for the hosts of horticulturists that are to follow. There is no place in the country that furnishes greater variety of soils than the rough or mountain lands. Often showing several distinct kinds of soil on one hillside, and where such is the case it may pay to change varieties accordingly, even in the same row.

ALTITUDE.—Among the many advantages of rolling lands for orcharding are the varied altitudes from which to choose. The grower may humor his own likes or dislikes in regard to varieties he may be desirous of planting. For instance, if he should wish to grow the Baldwin, Northern Spy and Greening, he should choose the higher altitude, approaching as nearly as possible the same climatic conditions under which we find these varieties succeeding. (Say New York conditions.)

If some of the varieties which succeed in more southern sections and consequently longer seasons should suit his fancies, then he should choose his location in a section of lower altitude.

Again the season of ripening of the same varieties of fruit may be much prolonged by the difference in the altitude of the plantations. For example, it is possible, by taking advantage of the varying altitudes, to gather the same varieties of apples or peaches over a much longer season. The varying altitudes not only give a fruit region a larger range of varieties which may be grown successfully, but also prolong the ripening season of the same varieties. Thereby avoiding competition in the home markets from the various parts of the same sections, while the consumer who lives in such a district is fortunate in that he may have the fresh fruit of the same variety for a longer season than those dependent upon more level sections for the production of their fruit supply.

"Some sing of the yellow apple
That grows so large and fair,
They tell of its many qualities

Which are seemingly so rare.
And yet, I have been thinking,
  That no matter what's been said
Tho' they may choose the yellow fruit
  That I prefer the red.

In the early part of summer,
  How pleasant 'tis to me,
To see the bright red apples
  All growing on the tree.
For 'tis when we have the 'Duchess'
  And the 'Astrachan' so red,
How can any one, I wonder,
  Choose the yellow fruit instead?

The Strawberry and the Snow apple
  Come later in the fall;
Oh, yes, you'd better try them
  For they're sure to please you all.
And there are many others
  Whose praises we could sing,
The famous 'Seek-no-further'
  And the 'Tompkins County King.'

So now you see the reason
  Why, as before I said,
Tho' they may choose the yellow fruit
  That I prefer the red.
And if you make the red your choice
  You need not have a fear,
The 'Wagner' and 'Ben Davis'
  Will last through all the year.

When you're in search of fruit trees,
  Remember what's been said,
And although you plant the yellow fruit
  Pray don't forget the red."

               —*V. M. Vose.*

A WELL DRAINED ORCHARD SITE
Courtesy J. H. Hutchinson

CHAPTER IV.
DRAINAGE.

"Over the orchard the raindrops fall,
Playing hide-and-seek round the old stone wall,
Sparkling like gems in the sunbeam's bright,
Kissing the daisies with lips so light.
Chasing the shadows that come and go,
As the dew laden trees nod to and fro,
Then, as if tired with the morning's play,
They run to the brook, and are swept away,
Whirling and rushing with never a rest,
Till clasped to their dear Mother Ocean's breast."
—*Will F. Stephens.*

There is no one thing that should receive more careful attention when selecting the orchard site than the subject of drainage. When we speak of drainage when applied to our farm crops, we naturally think of the draining off of the surplus water, in order that our crops may not be flooded or drowned out There are few operations which are so generally recognized as being essential to good farming, and few, if any, that pay a larger percent on the amount invested than drainage.

We all recognize the fact that roots of plants must have air as well as water in order to promote growth. How quickly corn will turn yellow in the wet spots over the field. It is not necessarily because of the lack of fertility in the soil, but it may be, and frequently is, because the air has been excluded from the soil by the excess of water.

We should recognize the difference between a moist and a water soaked soil. While moisture is necessary to the growing plant, on the other hand we cannot expect the plant to even live, much less grow, in a soil which is water soaked.

If it has paid to drain lands for a crop that may be renewed each season, and it has, (as some of our great swamp sections stand ready to testify by the production of their enormous crops, where but a few years ago only swamp grass and cat-tails flourished,) then it will certainly pay to see that our orchard sites, where we expect the trees to grow for years, are properly drained.

At first thought one might say, what has drainage to do with orcharding on rough lands? There are a great many places, even in rough or hilly sections, where we find trees dying from the effects of wet feet. It frequently happens that when the timber is removed from these hills that a great many spouty places, or wet weather springs as they are commonly called, make their appearance. If they are not taken care of by means of some sort, the trees will not only suffer, but landslides often occur, carrying the trees with them, and they sometimes get things, such as fruit trees, stumps, boulders, etc., very badly mixed. While a line of tile could have been laid which would have carried off the surplus water, and prevented its gathering in some low places and soaking down to the sub-soil, or to a ledge of rock, and thereby causing a slip. Such cases, while not common, sometimes occur, and the damage is very great.

A site that is naturally rolling, with a subsoil which is open enough to admit of a free passage of water, is the site that would be preferable, as the trees will root more deeply, live longer, and consequently be more profitable, than on hard-pan soils. As has already been said, plants and especially trees, do not reach their highest degree of perfection in wet soil Also we find that some varieties suffer more severely in this respect than others, for instance, the Grimes Golden seems to suffer from the trouble called (for want of a better name)

Root-Rot. This is particularly noticeable in orchards where water stands around the trees in the winter or during the early spring. Ben Davis suffers occasionally in the same way, while the Spy and Baldwin seem to be able to stand more water without damage than many others.

Root-rot may be recognized by the bark around the crown of the tree turning dark and coming loose from the wood, and when examined there will be found a slimy substance under the bark. This trouble may appear only on one side of the tree. The tree may linger along for a season or two, but sooner or later will succumb.

There are many other reasons why wet soil should be avoided, such as a severe freezing of the soil and roots. The difficulty in getting and retaining cover crops, such as clover, on account of the lifting of the plants during the freezing and thawing weather. Many orchards have been failures because the soil has been too wet and the planter has not considered the drainage of enough importance to warrant the necessary outlay of time and money. Again it has often been because of the thoughtlessness of the planter, never having had his attention called to the importance of drainage.

Surely those who want to engage in tree planting will have too much love for their trees to plant them where they can only linger and die. We should remember trees cannot move about, but must remain where we put them. Then we should see to it that the soil in which we expect them to grow and become a source of pleasure and profit is at least well drained, thereby making it favorable to their growth in this respect at least.

AIR OR FROST DRAINAGE.—Among the advantages of rough or rolling land for orcharding there is none of greater value to the fruit grower than air or frost drainage, as it offers us protection from late spring frosts. These conditions are more marked some seasons than others. However, the difference between different locations may be noticed almost any spring to a certain extent. We see the effect of frost drainage frequently even in our pasture fields, when we get out early some morning and find the grass frosty and even frozen on the low lands, while on the higher ground there is not the slightest trace of frost. We have often seen the leaves and even the young shoots of the hickory and pawpaw killed in the hollow and on the hillsides up to a certain place, while from there up, all would be green. Sometimes

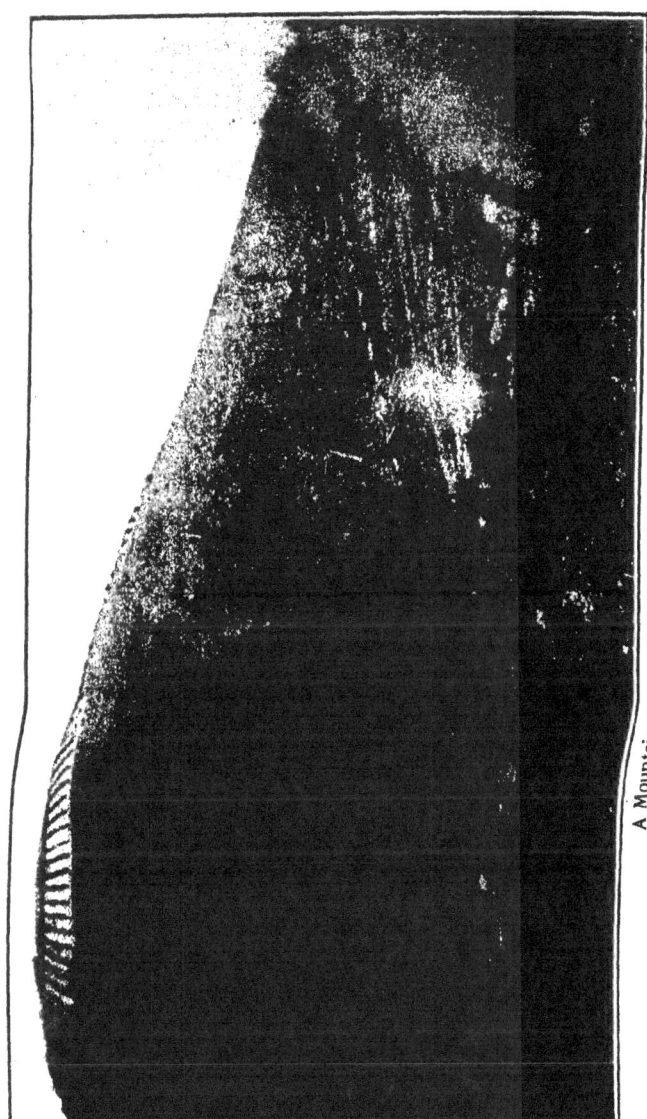

A Mountain Orchard With Splendid Air or Frost Drainage

we see a whole mountain side covered with a peach orchard, and a late spring frost will kill all the fruit for a certain distance up the mountain side, the rest of the trees being loaded with fruit. This is sometimes so marked that there may be seen a well defined line just at a certain level where the work of the frost has ceased.

We are reminded of frost or air drainage frequently when driving through the country after nightfall, and notice how much cooler it is at the mouth of a hollow than it was just before we came to it. All these instances show how the cold air (which is heavier than the warm air) is settling to the low lands, or draining off just as water would.

Now we would not be misunderstood in the matter of air drainage and leave the impression that altitude is the necessary factor. An orchard site may be on the top of a mountain at an elevation of 2000 feet above the sea level, and yet not have good air or frost drainage.

Again it is possible that an orchard site may be located at an elevation of only 500 feet above the sea level, and yet escape the frost more frequently than the one at an elevation of 2000 feet. Because in the case of the one on the high mountain, it might be situated on a large level plateau of several thousand acres, and there may even be slight depressions in the land, which would tend to act as frost pockets. While on the other hand the site which is only about 500 feet above the sea level is probably located on the top of the hill or ridge, with an abrupt descent on one or more sides, thus enabling the cold air to escape or be drawn off to the lower lands. So when considering air or frost drainage let us look for abrupt descents and not high elevations, always remembering that it is the comparative elevation of the surrounding lands and not the altitude of the site itself that counts in the matter of frost drainage.

We should note the difference between a frost and a freeze. Frosts occur on still, clear nights. We often say it will frost tonight if it should clear off and the wind stop blowing. Then frosts are more or less local; while freezes frequently come with or on the heels of a storm, so are not infrequently accompanied by a high wind and very often occur on cloudy nights. A freeze is more general over the whole section and will not play as many pranks of skip, hop and jump as does the frost. Then Jack Frost always wears a white coat.

Air drainage may protect against frost but not against a freeze, such as swept over many

Orchard at Foot of Mountain Which Suffers From Frost, on Account of Lack of Air-Drainage
Courtesy H. P. Gould

fruit sections the season of 1910, as late as the sixth of May.

There are other things that may enter into the matter of frost protection, such as large bodies of water near the orchard site. Then the direction of the wind from such bodies, whether toward or from the orchard. But these do not come up for consideration in the discussion of orcharding on rough lands. Here we have only to deal with air drainage, the value of which is being recognized more of late years than formerly.

Frost drainage, together with a more careful study of the various aspects afforded by our rolling lands, will enable us to take advantage to a certain extent, of the uncertain weather conditions, and thereby avoid many failures in seasons of late frosts.

The successful fruit grower of the future will be the man who has fruit when his neighbors' orchards fail, and in order to do this we must grow our fruit and not allow it simply to grow itself. Then let us choose carefully not only the site, but consider thoughtfully the aspect as well. By so doing we may take advantage of some of the opportunities afforded by our rough and rolling lands.

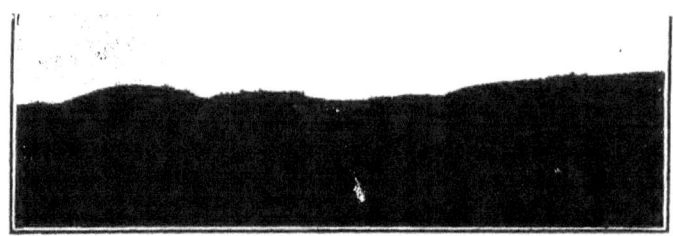
ORCHARDING WHERE CHOICE OF SLOPES MAY BE HAD
Courtesy R. L. Hutchinson

CHAPTER V.
THE ASPECT.

"Under the cloudless blue they lie,
  Golden hills in the golden sun;
Rising up to the mountains high,
  Reaching down where the rivers run.
Up to the springs of youth they lead,
  Under the edge of the purple pines;
Ways untainted by toil or greed,
  Paths where peace in its fulness shines."
—*Mabel Earle.*

The aspect of the orchard is a subject upon which there has been a great deal both said and written. A very great difference of opinion exists as to the advantage and disadvantage of the aspect or slope of the land upon which the orchard is planted. Some prefer a Northern, others a Southern, many choose the Eastern and quite frequently we find paying orchards which have a decided Western exposure. So it becomes a question as to whether there is any real difference, and if so what should govern us in our choice. We have found in our experience and observation that no one slope or exposure proves the best under all circumstances. The surroundings of the particular location, together with the varieties of fruits to be grown, with special reference to the demands of the markets for which it is to be produced, should largely govern us. Whether it be a demand for early, high-colored fruit, or a late maturing variety.

For example, in our orchard, the highest colored (and it naturally follows, the best quality) dessert apples have been grown on a Southern or a Southeastern slope, while the most bushels per tree during single seasons have perhaps been gathered from the more Northern exposures. However, these trees did not come into bearing as early in life as those on the Southern and Eastern slopes, and had consequently grown much larger. That exposure counts for much, especially to the grower of early, tender fruits, none will deny.

In order that we may understand more fully the reasons for these varying results, consequently choice or rejection of certain aspects, we should consider some of the advantages that may be gained in the choice of the aspect of the orchard. It is well to give this a little space, especially as we are studying orcharding on

Orchards With Various Aspects and Good Air Drainage
Courtesy H. P. Gould

rough land, where we may have choice of the various slopes, for it is hard to find any two of them that will be exactly alike when we consider them from the following viewpoints.

First, we should consider the strength of the soil, for after all that is the foundation upon which we must build. As a rule we find certain hillsides in some localities or parts of the country much more fertile than the opposite sides of the same hill or mountain. We have in mind timbered sections. This can generally be accounted for by the direction from which the prevailing winds have blown. For example in our own experience, we have found the Northern and Northeastern slopes much richer than either the Western or Southern, as the prevailing winds in the section in which these plantings were made are from the West and Southwest, so that for years and years the leaves had been blown over to the North sides of the hills. By their decay they have made that side the richest; while the Southern and Western sides of the same hill have been robbed (by the wind) of their just share of fertility by this constant loss of leaves.

Next let us consider the matter of moisture holding capacity of the two soils, for upon the moisture content of the soil largely depends its power of production. We all recognize the fact that without moisture the plant food (no matter how much there might be in the soil) could be of no use, so that moisture becomes an absolute necessity to the growth of plants.

When we speak of rich and poor soils we should not lose sight of the fact that often times the one we call poor may be just as rich as the other in its stores of plant food, but this plant food is not available. For instance, take the great desert, as we think of and call those immense tracts of lands west of the Rockies, which were made productive simply by the addition of moisture.

We should think of the availability of the plant food in the soil rather than of it as a rich or poor piece of ground. We prefer to say unproductive soil rather than poor or worn-out soil, as it is so often expressed. The one thing that counts for more, not only in the retention of moisture, but in setting free the plant food that is already in the soil, as well as adding to it, is decaying vegetation, or as we say, humus. This is why we find some of our hillsides, as have been mentioned, the richer because of the constant drifting of the forest leaves, the decay of which has not only added plant food but has increased the humus as well. Humus acts as a sponge that takes up and holds the moisture for the use of the plants. There may be orchards where there is an excess of moisture, or too much at the wrong time, thereby causing wood growth to be made at the expense of the formation of fruit buds. For this reason we find trees continuing to grow on the North hillsides long after those on the South have ceased. Again we often find the trees much larger at a given age, and just as frequently they are late coming into bearing. This will be discussed more fully under the head of Cultivation.

THE WARMTH OF THE SOIL.--The warmth of the soil will be governed; first, by the character of the soil; and secondly, by the slope of the land. Where the soil has a great deal of sand or gravel in it, we call it a warm or early soil. A warm soil is especially valuable in the growing of strawberries, early fruits and vegetables, but has likely claimed more attention from the berry growers than any other class of horticulturists up to the present time. When we wish to produce the extra early varieties of berries, we look for Southern slopes with a warm soil. If we want to furnish a late market we use a late variety and choose a Northern slope, thus recognizing the difference in the warmth of the soil as determined principally by the aspect.

PROTECTION FROM WINDS.—The protection of fruit plantations from heavy winds is another phase of the question that should not be overlooked when considering the aspect, as it may mean much to the grower. For instance, a Western slope may be so much exposed to constant winds that the trees may suffer by being blown or worked around during soft, open weather in winter. Then the snows will be blown off much worse, thereby depriving the soil of a supply of moisture which it would have received from the melting snow. Not only has the soil lost the moisture supply, but the cover crops, if there be any, have lost the protection that the snows would have afforded them, and the soil is frequently left bare during long cold spells which follow windy snow storms. Sometimes the roots of the trees suffer from freezing.

Again we find that the leaves are all blown from our orchards. This is a loss of fertility; or they are drifted into the low places, and this is a source of danger when allowed to pile around the trees, as it is almost impossible to keep the mice from harboring in them and barking the trees. Frequently, even when the bodies of the trees are protected we find the mice burrowing down under the leaves and peeling the entire root system.

Sweeping winds not only carry off the moisture from the soil, but our trees often suffer severely from the loss of moisture given off from their branches and twigs. This is frequently over looked by the planter, and after setting young trees in the fall on very windy exposures, he is surprised to find when spring comes that many of them are nearly or quite killed to the ground. Especially if the trees are young, (say yearlings), as the moisture may be given off more freely from trees with young bark than those whose bark has become thick and heavy.

This is discussed more fully under the head of Wind-breaks.

ORCHARD WHERE TIMBER SERVED AS A WINDBREAK
Courtesy W. E. Rumsey

CHAPTER VI.
WINDBREAKS.

"In patient, silent ranks they stand, a wall
 Of purple shadow 'gainst the sky's dull gray,
Not dead, but only dreaming of the day
 That once again shall voice the sweet spring's call."

There used to be almost as many differences of opinion as to the advantages and disadvantages of windbreaks as there were different orchardists. Much emphasis has been placed upon the subject, and large expenditures of money have been made to grow windbreaks. But of late years the orchardists do not consider windbreaks of as great importance as many other things which are to be considered in the location of orchards.

In level sections where there are almost constant sweeping winds they might be a means of protection, both to the trees while growing and to the fruit, thus lessening the windfalls.

Or in holding the leaves on the ground in the fall, and the snow in the winter, as a protection to the roots of the trees, as well as to afford protection to any cover crop that might be growing in the orchard.

Windbreaks might afford some protection for the work of spraying; also be worth while for heavily loaded trees, thereby lessening the damage that often occurs by the breaking of limbs in severe storms.

These are some of the arguments that we have used when trying to convince ourselves that artificial windbreaks were worth while, and we believe under certain conditions they may be useful. On the other hand there are objections that to our mind are well founded, and what we shall say in regard to this, as well as along all other lines, will be from our own practical experience.

We planted an orchard of twelve acres in the spring of 1896. The location was a front river hill, with a level stretch of bottom land a mile wide between it and the river which was directly west of the site.

The site was a large cove opened to the east. This opening was almost at the half way place in the east line of the site, also being the lowest point. The land then rose rather steeply from this point, or opening, in three directions, making three distinct exposures; one to the south, one to the east and one to the north. All these slopes culminating at the top of rather a sharp hill or bluff, which formed the rim of the cove. This rim, so to speak, is timbered on the entire west side of the site, the timber being a second growth of black oak, chestnut oak, ash and black locust. The timber grew from the crest of the rim over the entire west face of the hill, which is probably 200 feet wide.

Having some faith in artificial windbreaks we were glad to find ourselves so well supplied with this one placed there by nature. The trees were all removed from the top of the rim far enough to admit of the planting of the first row of apple trees right on the crest. This left the tops of the remaining trees in the windbreak 20 to 30 feet higher than the crest of the site. Conditions seemed ideal, but results have not been. There has been greater losses by late frosts directly under the shelter of the windbreak (owing to the lack of circulation of air) than there has been even at the lowest point of the site.

We all know how much the wind is broken by belts of timber. We frequently see stock seeking such sheltered places during storms.

This is because the wind rises when it strikes the timber, and thus leaves a quiet space, or as we say, a dead air space next to the timber. It is in such places as this that the frost frequently does its most harmful work. While a little distance from the windbreak, where the air is in motion, the fruit will likely escape damage.

Then we have not been able to get the wood growth in the first row of trees that we have in the remainder of the orchard, although the roots of the trees which form the windbreak are down on the hillside many feet below those of the apple trees and should not interfere with them, but this row seems to suffer from droughts much more than any others in the orchard. There has been more poorly colored fruit against this windbreak than in any other portion of the orchard, and fungous troubles have been hard to control in the lee of the trees. The leaves from the windbreak are blown over this ridge and lie in great drifts around the first two rows of trees. It has proven almost a hopeless task to prevent the mice from breeding among the roots of the trees of the windbreak. Then during the winter they work under the leaves to the fruit trees and bark them, frequently killing trees that have reached bearing age.

Orchard Where Timber Covered Mountain Served as Wind Break
Courtesy H. P. Gould

While the orchard has had some protection from severe storms, and probably the windfalls have been lessened, we are satisfied that the advantages have been overbalanced by the disadvantages. If you want the advantages of windbreaks you can find most desirable ones already for use on the rough lands, by choosing locations where advantages may be taken of some still higher hills. Or hills of practically the same height whose covering of timber has not been removed, even if this hill is some distance from the orchard site, you will see the protecting effect. In fact, we should prefer that it be slightly removed, as our experience and observation has taught us that without a fairly good circulation of air in the orchard it is hard to keep down many of our troubles, even with a spray pump, and we have concluded that it is better to have windfalls than wormfalls.

From experience we have decided that it is better to build the orchard so it will be able to stand the winds of the section in which it is to grow, excepting, of course, such storms as uproot the mighty oak. The way we have tried to do this is first by planting closely, so that each tree may help to protect its neighbor. Then try to avoid alley ways for the wind to sweep through, by not allowing the rows to run the same direction of the prevailing winds.

We prefer to have the prevailing winds strike the rows cross-wise of the broad middles, that is, when any difference is made in their width. Occasionally we find orchards laid off in blocks with wide driveways between, and these drive ways prove (as the orchards grow larger) to be wind-sweeps through the plantation. If they had been laid off with this in mind, and been allowed to run so that the prevailing winds would have blown across rather than with them, the good and comfort that they were intended to afford might have been enjoyed without this harmful effect.

We should not only plant for protection, but we should prune the trees so as to form low heads. Low enough that the wind would pass over rather than under them. The heads should be left more dense than is often practiced in more sheltered localities, so that each branch, in time of storm, may to a certain extent protect the other. In pruning, make the cuts short, so that the branches will not be long enough for the wind to thrash them against the ground, or break them by swaying back and forth with their heavy load.

We believe if these points are kept in mind

72   *Practical Orcharding On Rough Lands.*

that we may grow good orchards and reap abundant harvests without any other windbreaks than are afforded by the natural contour of the rough and rolling lands, if we have this in mind when choosing our site and make our choice accordingly.

Orchard Site Where Stumps Have Been Removed.
Courtesy G. T. Leatherman

PREPARING THE ORCHARD SITE
Courtesy E. R. Lake

CHAPTER VII.

PREPARATION OF THE SITE.

"The dark brown earth's upturned,
By the sharp pointed plough—
And I've a lesson learned.

My life is but a field
Stretched out beneath God's sky,
Some harvest rich to yield."

The preparation of the orchard site is a matter which should and must have more thought and study by the future fruit growers if they expect to succeed, and not as it frequently is, be ignored and treated lightly. The average corn grower of today prepares his seed bed most thoroughly, although the crop to be planted will only occupy it for a few months. Then the land may be refitted, and sown or planted to something else.

If careful preparation is necessary for quick maturing crops, how much more careful should we be in the preparation of the orchard site, for it is only once we shall have the opportunity to prepare this land. After the trees are in place we can only ·cultivate the surface, so the preparation for tree planting should be most thorough.

The preparation of an orchard site is a task that should be a source of great pleasure to the tree lover, for he should have in mind that he is beginning a work, laying the foundation, if you please, for a building, the beauty and usefulness of which will largely depend upon, as in other structures, the thoroughness with which the work is done.

While it is important that every detail be considered carefully in the preparation of an orchard site, this work, like a great many other things the orchardist is called upon to do, cannot be done according to any set rules. Every orchard site may present to the planter an entirely different problem The contour of the land may be the first factor to be considered, for if it is fairly level you might proceed one way, while if the site is steep an entirely different line of preparation would be followed. If it should be an old worn field, then you could plow thoroughly and might even use a sub-soiler to an advantage. While on the other hand, if it is new land just cleared, the method suggested for the old field would not suit at all. So we should study our conditions and choose our methods accordingly. In the preparation of woodlands the question may be asked as to whether the timber should be removed by grubbing, or simply by chopping it off. Here, as in many other farm preparations we find a very great difference of opinion. We shall not try to settle this or any other disputed point. But let us keep in mind the object in view, namely, getting rid of the brush, roots and stumps of the forest, and preparing the land for the roots of the fruit trees. Any method that will accomplish this in the shortest time and with the least labor is the one that should be adopted or practiced.

In our experience we find that we can accomplish this quicker and cheaper by chopping off than by grubbing. The explanation we have to offer is that when we grub a bush or tree, we likely remove the first six or ten inches of the growth below the ground. Then the stump (for that is what it is), of the root sends up sprouts, which are most likely cut off at or near the surface, thus leaving a stub below the ground. This stub sprouts, and is probably cut off in the same manner. This being repeated time after time, forms a stool just below the surface that is much harder to kill than the stump above the ground would have been. When sprouts are removed from stumps there are no new stools allowed to form. Then when decay begins above the ground it soon extends to the root system, and the stump is not only dead, but out and gone before you have been able to kill the sprouts and stools in the land that was grubbed.

In preparing the orchard site we should remember the one object is to enable the roots of the trees to occupy the entire area. It is best when possible to plow the entire surface, although many thousands of trees are being set on rough mountain land, with only a few furrows thrown out for the tree row, and the remainder of the surface to be broken up later. When this is done there should be just as thorough preparation of the place where the tree is to stand as possible. This practice may be satisfactory on land that has an abundant supply of stone or gravel in it, or land that has a very porous subsoil, so that there would

An Orchard Site in Winter, Prepared for Spring Setting
Courtesy Leo Jellinek

be no danger of washing. It would be somewhat dangerous on clay lands that were rolling, as the water is likely to gather in the furrows made for the tree rows, and then when it breaks over, as it is sure to do, it will cause washing which we can never afford to allow.

The orchard site should be thoroughly plowed in the fall or early winter, so as to get the action of the frost on the soil, as well as to hasten the work in the spring. The tree holes should be dug as early as possible in order that the sub-soil should be allowed to freeze during the winter. This freezing and thawing loosens up the sub-soil in a way that nothing else will, as well as allowing the water to soak into it which acts as a reservoir the following season in supplying moisture for the young tree. This will not be necessary on gravelly or porous soils. On such land the holes may be opened at planting time with perfect safety. The size of the holes should be governed by the kind of soil in which you are planting. If it is loose and rich, then a medium size hole will be sufficient. If you are planting in thin, heavy soils, then you will be well repaid by a better growth, for the extra labor in digging the larger hole, say three feet square and eighteen inches deep.

> "Dig three feet deep, each planting hole,
> Fix 'plumb' each stem with twine and pole,
> The loose sward underneath you cast,
> Fill up with earth and tread it fast;
> Then leave your tree, but don't forget
> How much the loosened soil will set,
> To catch the rain the earth about
> Should like a dish be hollowed out."

There are some sites which are so very steep that the plow can not be used to advantage in their preparation. If these soils are filled with stone or gravel so that the roots of the trees may penetrate it easily the planter may succeed by digging the holes without plowing the surface at all. This method will require more hand labor (or digging around the trees) than if the surface had been plowed. This plan does not succeed so well in tight clay soils, as the water frequently gathers in the holes on account of lack of drainage.

It will be found advantageous when digging the holes to pile the top soil to itself, so it may be at hand when setting the trees. On steep land it has proven to be a good plan to pile the dirt on the upper side of the hole, so that the washings may be caught in the hole below. We should not lose sight of the fact that it is the lightest and best of the soil that is carried away by the washing, (erosion), and when once outside the boundaries of our deed it is gone forever, so we should be very careful about its loss.

The matter of drainage has been spoken of, but we wish to call attention to its importance again. After the holes are dug in the fall or early winter, so that we may have the benefit of the action of the frost which has already been described, they should be visited occasionally during the winter. If the water is found to be standing in any of them, drainage should be provided; if by no other means than by sinking another hole on the lower side and deeper than the one intended for the tree, and the water drained into it. This will prove all the more effective if the hole is filled with stones, stumps, or rubbish of some kind before it is covered over or filled with dirt. This insures its remaining open until the trees become established, after which the tree roots tend to act as conductors, which prevent the collecting of water about the bodies of the trees.

Much valuable time may be saved in the rush of spring by this winter work. Besides we are able to plant much earlier where the holes are ready, and the dirt—if it is piled where it will drain properly—will be in condition to work several days before that of the level surface. It is even more important to have the ground in good condition for tree planting than for ordinary farm crops. A thorough preparation of the entire surface will eventually be necessary in order to get the best results, unless it should be in very open, stony soil, as already described.

The fact that we figure as to how many trees can be planted to the acre without crowding should be enough to convince us of a fact which so few seem to realize, that we expect the roots to occupy the entire surface. If so, then we should see to it that the soil is prepared in such a manner that the trees may be able to take entire possession.

THE OLD PLOW.

"By the fence in the orchard the old plow stands,
   Slowly rusting and rotting away,
While the days go by with their dropping sands
   And the world grows dull and gray.

It did its work in the long ago
   As it tumbled the stony soil,
And the harvest waved with a golden glow
   With a crown for the brow of toil.

It seemed to shout like a warrior bold
   As it entered the stubborn field,
And the wind-swept clouds above it rolled
   And the sun smote its shining shield.

But now it stands by the fence alone,
   With its share all brown with rust,
And its oaken frame with weeds o'er grown
   And mouldering away to dust.

And soon I know with the flowing tide
   That furrows the silvered brow,
I, too, will be tenderly laid aside
   To moulder like an out-worn plow."

—*Rev. George Crofts.*

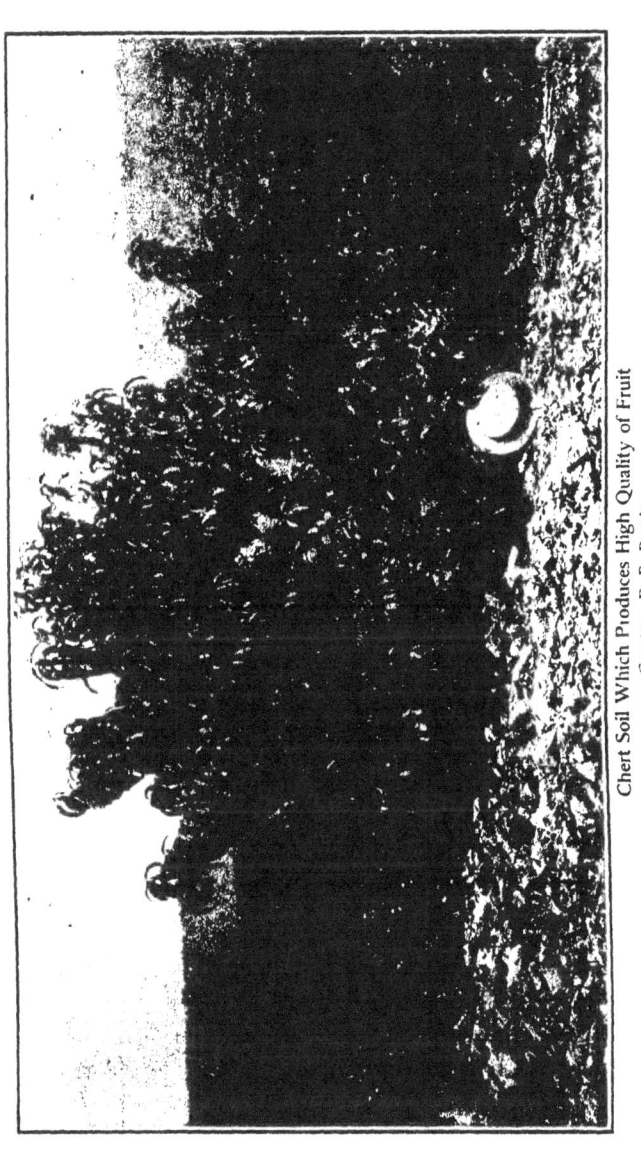

Chert Soil Which Produces High Quality of Fruit
Courtesy F. E. Brooks

AWAITING SPRING
Courtesy Leo Jellinek

CHAPTER VIII.

LAYING OFF THE ORCHARD.

"Be careful not to plant too close,
Your profit not from numbers flows,
Your trees require both light and space
That they may grow and thrive apace;
Set them, well, thirty feet apart—
In rows arranged with studious art;
For years to come the orchard will
Attest the planter's taste and skill."

Laying off the orchard is one of the operations that should be well outlined in the planter's mind before the work is commenced. There are many things that should be taken into consideration. The contour of the land should be studied, as it is necessary that the rows be made so that when the time of spraying arrives the planter may be able to drive over the orchard with the greatest possible ease. The direction from which the prevailing winds may come should govern the course of the rows to a certain extent, as we prefer to have the wind blow across them the narrow way, when there is any difference made in the space between the rows, (as described under windbreaks). In cultivation it will be necessary to follow the rows, and at the same time endeavor to make all furrows in such a way as to prevent washing as much as possible. This would not have to claim the attention of the planter when deciding upon the plan of laying off his orchard if his land were level, but when orcharding on rough and rolling land we become more and more impressed with its importance.

The plan of setting should next be decided upon. There are two general plans, the old square and the new triangular. The latter plan has grown in favor very much of late among planters, especially on rolling lands. It is a plan which enables the orchardist to take advantage of the contour of the land, as much as any we have practiced. This plan places the trees of one row opposite the space in the next row.

In the square form every tree stands in the corner of a square and equally distant from four others, while in the hit and miss, or triangular, every tree stands in the angle of a triangle of equal sides and in the center of and equally distant from six others. So in the latter plan there is a greater space left for the admission of light and air, although the trees are planted at a less distance than in the other plan.

The difference in the number of trees which may be planted on an acre under the two methods, the old and the new,—the square and the triangular—is shown by the following table:

```
                                          Old     New
Trees planted 30 ft. apart each way,   49 to acre  59
Trees planted 24 ft. apart each way,   75 to acre  90
Trees planted 20 ft. apart each way,  108 to acre 128
```

One great advantage to be gained by this triangular method, as shown by the table is the ability to grow more trees per acre, as well as more completely shading the ground, which should not be overlooked. The nearer we may approach forest conditions under our trees, and not interfere with the production of fruit, the nearer we have solved the problem of the cultivation of that much of the surface.

The operation of laying off the orchard in triangles may seem a little more complicated than in squares, but when we once get the plan clearly in mind we shall not have any trouble. We have found that this hit and miss, or triangular plan, may be best carried out in the following manner. After deciding how far apart the trees are to be set, cut two wires—No. 9 is a convenient size—the desired length, then fasten a ring on each end of both wires; this being done, we should next establish a base

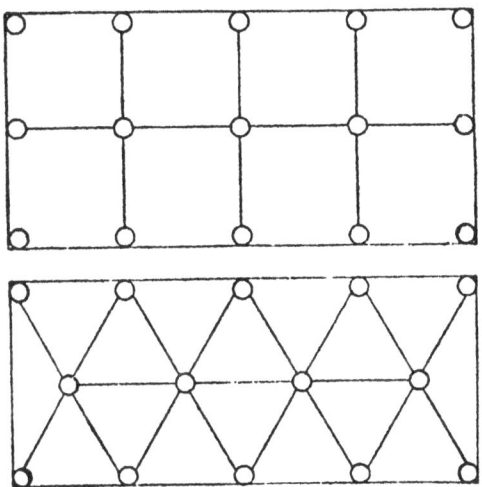

line from which to measure. Choose a straight side of the site, if there be one, if not, stake out a line. Then with a man at each end of a wire, with a good stout staff passed through the rings, measure off the distance desired between the trees by stretching the wire tight from staff to staff, always having them in line with the stakes which have been established. Each measure should be marked with a peg which will indicate the places for the trees in the first row or base line.

We are now ready to lay off the second row. Having our base line established, we shall now use both wires and the third man, who will take one end of both wires and pass the rings over his staff, the other men will stand each at a peg established in the other row, retaining one end of a wire, and holding it in place by the ring which passed over their staff. The third man now moves in a direction that will stretch both wires and when they are tight he places a peg at the place marked by his staff; this indicates the first tree in the second row. If the ground is steep and the third man is on the upper side of the hill, he should allow the rings to come nearly or quite to the ground on his staff, while the men who are at the pegs established in the base line, should slip their rings up on their staff far enough to approach as nearly as possible a level. If the man starting the new row is down hill from those on the base line, he should slip his rings—and they should be fastened together—up while the men in the other row should lower theirs, the nearer the wire is kept on a level the truer or more exact will be the rows.

After the first tree in the second row is

An Orchard Apparently Without Any System in Regard to Manner of Setting
Courtesy H. P. Gould

located all move back along the base line one peg; this places the man who stood at the end peg at the second one, where the other man stood who is now at the third peg in the base line; while the third man with the ends of both wires is tightening them again. When this is accomplished, and a level is approached a peg is placed at the base of his staff, when all move up as before, forming another angle, so to speak, with the two wires and the base line. When a row is thus completed it becomes a base line from which the next measurements are made. This plan will work over almost any kind of ground. Every operator should be careful to have his staff perpendicular when the wire is leveled and measurements made. For if they are held at an angle they will tend to shorten the distance between the trees, or get them out of line. Rows may be lengthened at any point in the plantation by making measurements just as in establishing the base line, using the pegs which indicate places for trees to site to as the stakes were used to establish the base line.

THE DISTANCE APART.—Having decided as to the plan of setting, the distance apart that the trees should be set would next claim consideration. This should be largely governed, first by the contour of the land, for the steeper the site the closer the planting may be made, as the rows are terraced, so to speak, one above the other. While if on a more level site it would be necessary to give them more room both for spread of branches as well as for better circulation of air.

The strength of the land should also have careful consideration. Thin, shaly soil could not be expected to give the growth that would be obtained on deep, rich soils. So if the land should be thin and steep the trees may be planted much closer than on the rich level site. It should be the aim of the planter to grow just as many trees per acre as possible, so long as they will not interfere with each other's development. Remembering that the more trees we have on a given area the more divided the risk, consequently the more chance for profit, with practically the same expense of cultivation per acre.

Varieties and their habits of growth should have an important part to play in the decision of this question. The casual observer cannot fail to recognize these characteristics. For example, look at the yellow Transparent with its upright top, almost as much so as the pear; then the Mammoth Black Twig with a great open sprawling head, the one requiring almost as much more space as the other. Shading of the land should be considered when deciding upon the distance for setting the trees. This is of special importance on many of the steep shaly southern exposures which are being planted to orchards, as we find it is absolutely necessary to shade these lands in order to get best results. This will be more fully explained under the head of Pruning. This may be best accomplished by the plan of setting the trees hit or miss as already described.

FILLERS.—The practice of setting early bearing trees between the permanent ones as fillers has been practiced largely in some sections of the country, and has some features to recommend it. Especially when using high priced land, or when the grower is in need of early returns to help develop the plantation. We have known of a case where peach trees were used as fillers in the apple orchard, there being two peach trees set for each apple. The rows of apple, thirty-two feet apart, then a row of peach between them one way, setting the peach sixteen feet apart. When this orchard was six years old the peach had borne three crops and had paid the whole expense of growing the apple trees and the peach orchard up to that time. When the peaches were cut out at six

An Orchard Where Fillers Were Used
Courtesy Leo Jellinek

years of age the apple orchard was turned over to the grower absolutely free. So it would seem that the planting of fillers should be recommended for general practice. But our experience and observation would not warrant recommending the practice generally. First, as land is cheap, there is no excuse for crowding the plantings. Secondly, we can come nearer choosing soil that is suited to one rather than suited to both apples or peaches. Third, the cultivation required in order to succeed in each is so very different. Then again when we come to spraying, we find it almost impossible to make the applications to the one without allowing some of the mixture to fall on the other. It often happens that it is necessary to use a mixture for the apple that would be entirely too strong for the foliage of the peach, and we can never afford to do anything that injures the foliage of our trees.

We are apt to allow the fillers to remain too long and thereby injure the permanent trees. It is after a test of this kind that we wish to emphasize how hard it is to go in and cut out good healthy bearing trees, even if we do think there may be danger of checking or stunting the growth of the orchard proper by leaving them. This they could easily do to such an extent that the future usefulness of the orchard would be damaged more than all the crops of peaches had amounted to, while if they had been taken out at the right time, although it might have seemed to be a sacrifice, the peaches would have proven a source of profit. The hardest days orcharding we ever did was to cut out 1500 fine, healthy six year old peach trees which had paid for the whole planting both of apple and peach. But it proved to be a profitable operation, for today it is a fine apple orchard which has, and promises to be a source of profit for years. A neighbor left his fillers in, and neither the peach nor apple give any promise of being a profitable investment, although the one plantation was as promising as the other in their early years.

If the planter feels that he MUST use fillers, then use apples, choosing some of the early bearing sorts which have an upright habit of growth, such as Transparent and Wealthy. The latter, however, is more of a spreading head, but a good early bearer. The great advantage of the more upright growers is that they may remain longer without crowding. By using apples as fillers it will enable the planter to cultivate and spray the whole orchard without interfering with the growth, or running any risk from the use of sprays, as might be the case if peaches were planted, as their cultivation should differ from the apple. No matter what is used, we should not forget the great danger of allowing them to remain too long. For, as Judge Samuel Miller, of Bluffton, Mo., so well said, when asked his advice upon this subject, "Fillers may be a good thing in the hands of the grower who has the nerve to take them out at the right time." There is more in this advice than the inexperienced fruit grower may think. Another point which should not be overlooked is the proportionate expense of growing the fillers as compared with their period of usefulness. The first cost and early care of the trees will be as great as though they were to be permanent. This is of special importance on rough land, as we cannot economize in labor as we could on more level lands. In view of all these facts we believe that on our rolling lands which are cheap as compared with land in many other fruit sections it will prove more profitable to make the plantings permanent, even if we do have to wait a little longer for a large return per acre.

There may be orchard sites on steep hillsides which consist of series of benches, or narrow flats that run about the same level or altitude, on or around the hill. In some of these cases, where the driving with the spray tanks will all have to be done with strict regard to the contour of the land, it may be advisable to run the rows with the contour of the land or with the benches. Then all working and hauling may be governed accordingly.

Location or arrangement of varieties was mentioned under the head of planting. We should not forget, however, that a great deal may depend not only upon their arrangement in regard to the time of blooming, but great advantage may be gained in the matter of hauling, by not mixing varieties when loading, which causes confusion. We find it to be an advantage to begin loading on the high land and gather the packages as we go down the hill, the orchard should be laid off accordingly.

GROWING THE NURSERY STOCK
Courtesy M. Schwartzwalder

CHAPTER IX.

SELECTION AND CARE OF THE NURSERY STOCK.

"Be sure and choose a thrifty tree—
Let it not spoiled or stunted be;
Straight, tapering trunk; bark smooth and round,
Well shaped, fair sized the limbs around,
From top to base in prime condition;
Buy good home stock and save commission,
Plant few varieties, but choose
The very best—the rest refuse."

There should be no more important question that the prospective orchardist is called upon to consider than the selection of the Nursery Stock. Too many fail to realize how very important a matter this is, and frequently it receives little or no thought or study. In the past the average planter has paid but little attention to any of the various points that should be considered in the purchasing of trees except to try to get the cheapest that could be found. The successful fruit grower of the future will plant nothing but the very best regardless of the price. We have passed the time when we can afford to allow fruit to grow itself, and are living in the day when we must GROW the fruit if we expect to have it. In order to do this we must have a good tree.

WHERE TO PURCHASE.—Where to secure the proper kind of nursery stock is indeed a serious question. Generally speaking it will be best to get it from the nursery nearest home, thus obviating the long haul and at the same time lessening the danger of damage or possible loss of trees by being out of the ground so long. The home nurseryman will be more interested in the success of the planting than the man who never saw or expected to see the grower or the orchard. Another advantage in getting the trees near home is that the nursery may be visited and the trees selected. This means a slight expense, but we should be willing to go to some extra trouble and even expense in order to get the very best stock. The average farmer would not think of purchasing a lot of work horses or mules without seeing them. He expects them to serve him for three or four years, so is willing to go to considerable expense to select them.

This same careful business man will buy thousands of trees without seeing a single one of them. These trees are to be a source of profit or loss for years. He expects—or should expect —to bestow years of labor on them and yet he has not given them the thought or consideration that he did the work stock which could only last a few years at best.

Poor nursery stock has done more to retard the development of orcharding all over the country than any other one thing. The farmers and fruit growers are largely to blame for this. As has already been said they have been willing to buy the very cheapest trees that could be found. Not only have they been willing to buy, but they have demanded cheap nursery stock. We find in all lines of business whenever a cheap article is demanded some one is ready to furnish it. Nurserymen are no exception. In order that they may furnish the stock at the figure demanded and still realize a profit—and they would be foolish to attempt to do business without a profit—they simply put in enough trees that should have gone on the brush heap to enable them to fill the order. In this way the fruit growers are to blame for the poor trees they have received. They fail to realize that a poor tree might be dear as a gift, while a good one may prove a paying investment at what may seem at the time an extravagant price. It would be hard to say how much a planter could afford to pay for good trees rather than set poor ones.

For example, a good Rome Beauty tree after twelve years of care yielded $30.00 worth of fruit in a single season, besides some former crops. Alongside of this stood another of the same variety which was planted and cared for exactly as the other, yet it had never produced a single bushel. The one a good tree, the other diseased. The careful planter should not only figure as to the number of trees necessary to plant per acre, but should just as carefully consider the quality of those to be planted.

A GOOD TREE.—A good tree is one that has a good root system and should be well grown according to variety. Free from Insect Pests and Fungous troubles. The root system is important. By a good tree would be meant one with roots extending in all directions. The first purpose of the roots is to hold, or anchor the tree. This is shown by the acorn that is noticed among the leaves in the woods. Its first effort is to fasten or anchor itself to the ground by sending down the tap root. After which the tiny oak makes its appearance. Whenever trees have to be staked in order to keep them in place that orchard has many chances against it. When selecting nursery stock and examining the root system think of them first as anchors, and secondly as the means of gathering plant food from the soil. In order to succeed in either or both of these functions it is necessary that the roots extend in all directions.

SHOULD BE WELL GROWN ACCORDING TO VARIETY.—Planters are frequently to blame for the poor trees delivered to them. They have demanded trees of a certain size, regardless of the variety and their habit of growth. It should be kept in mind that all yearling trees are not the same size. For instance, a one year Stark may be as large as a two year Transparent. The habit of growth of the different varieties should always be taken into consideration. The orchardist who orders Stark, Transparent, Grimes, Jonathan, York and Rome Beauty two years old and expects or demands that they all be one size is not only making it possible but is almost forcing the nurseryman to put in a lot of three-year-old trees of the weaker growing varieties in order to have them near the same size, so as to please his customer. The difference in habit of growth of the different varieties should claim more attention of the prospective planter when choosing nursery stock. The small willowy Jonathan may be just as well grown according to its habits of growth, as the straight heavy Stark, which is much larger.

FREE FROM INSECTS.—There are many trees that are infested with scale when set, notwithstanding the fact that the States' Inspectors are as diligent as possible. It is almost an utter impossibility for every tree to be carefully examined. The time and money at the command of the inspectors will not permit of it. So the examination is necessarily general. Every planter should see that the trees received are free from scale of all kinds. Examine closely around the buds for the stragglers.

APHIS KNOTS.—The roots should be free from aphis knots which often cause them to look like strings of beads, when the tree is badly infested. In bending a root covered with these knots it frequently breaks as the constant puncturing of its tissues by the insect as it fed on the sap has made many wounds, thereby causing these knots. Although they had healed over, the fiber of the roots remained weakened. This should not be mistaken for or confused with some of the fungous troubles, as it is the effect of an insect, and not a disease. Trees

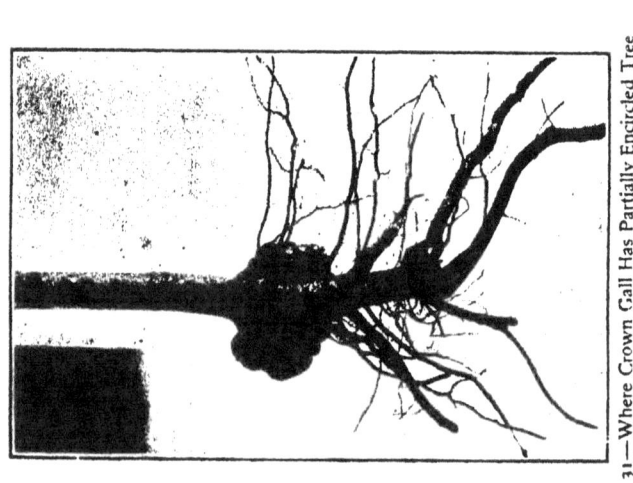

31—Where Crown Gall Has Partially Encircled Tree It May Live for Some Time

31—Where Crown Gall Has Entirely Surrounded Root and the Tree Will Die

Courtesy Dr. J. C. Whitten

thus affected should be discarded, or thoroughly treated with some contact insecticide. The safer plan would be not to set them, as it is very hard to eradicate the aphis by dipping the trees, which is a proposed remedy. They are frequently protected from the solutions by dried dirt on the roots.

FUNGOUS TROUBLES.—The principal fungous trouble that should be guarded against is Crown Gall. This may be recognized by a rough warty growth at or near the collar of the tree. It sometimes makes its appearance on the roots. These enlargements are frequently covered with fine hairy roots giving them a horse tail appearance. Whether these are the same thing or not will make but little difference to the planter, as he should discard and destroy all trees, which show signs of either of these troubles, Crown Gall or Hairy Root.

MANNER OF PROPAGATION.—There has been a great deal said and written upon the subject of propagation of trees, especially of the apple. This is not the time or place to discuss the relative value of Whole Root or Piece Root grafts; or to decide whether or not the Budded tree is superior to either. Suffice it to say that the orchardist who chooses a good tree because it has a good root system and is well grown, and discards all poor ones regardless of the method used in their propagation, is the man who is most likely to succeed. While the planter who sets a tree, good or bad, simply because it was propagated according to his pet method whether it be Budded, Whole Root or Piece Root grafted is likely to meet many disappointments, if not failures. Then choose the good tree regardless of the manner used in propagation.

GRIMES GOLDEN.—This variety should be grown on other stock rather than its own roots. We have found it a good plan to graft or bud it about one foot from the ground on some varieties like Northern Spy or Baldwin which do not seem to be subject to the Root Rot. In this way the bark of the Grimes will not come in contact with the soil and the trouble seems to be avoided. We believe the day is not far distant when we as orchardists shall pay more attention to the stock on which our trees are grown than to the manner used in their propagation.

WHAT AGED TREES TO PLANT.—Much will depend upon the kind of land that is to be used. If it is rough and rocky with roots and stumps to contend with, and if the labor is to be largely disinterested hired men, the two-year-old tree will likely be more satisfactory, as they may be seen somewhat easier than the yearling switch, among the sprouts and stumps. If the one-year tree is used it should have special care and cultivation the first season, or it may show the lack of its second year's treatment in the nursery, of which it has been deprived. One of the great advantages of the switch tree is the opportunity for each grower to train and shape the head as he may desire. Never plant an apple tree that is more than two years old. While some of them may live, the risk is too great, and the root system has been too long established to permit of a change without a serious check. The sooner a tree is in its permanent place,—all things being equal—the stronger will be its root system, and consequently the longer its period of usefulness.

COMPARATIVE HARDNESS IN FALL AND SPRING SETTING.—If fall planting is to be practiced and the site is exposed to constant winds the two-year-old tree will likely stand the winter better than the yearling. The former has had a better opportunity to ripen its wood, and has possibly more plant food stored in its roots, stems and branches than the switch. Its bark will probably be thicker and the moisture will not escape as readily as from the younger tree, which likely grew very late in the fall and failed to ripen perfectly either wood, bark or buds. When one year switches are used it would be well to set in the spring. They should be carefully protected during winter in trench or cellar, trench preferable.

TOO EARLY DELIVERY.—Fall is the time to secure nursery stock, even for spring setting. Care should be taken, however, that the delivery is not made too early. It sometimes happens that trees will be delivered to the farmer or fruit grower before the leaves have begun to fall from the trees in his home orchard. They would not be received by the most thoughtless grower if they were covered with dead or wilted leaves. But the average customer receives and pays for them without a word, although the bark may be very much shrivelled and the twigs quite dried out.

STRIPPING OF NURSERY TREES.—In order to make early deliveries the nurserymen often strip the leaves from the trees while they are standing in the row. This practice should be discouraged, or at least not encouraged by the planters insisting on early deliveries.

PURPOSE OF THE LEAF.—The purpose of the leaves should not be forgotten. They will complete their work under normal conditions, if allowed to. Unless they are permitted to remain on the tree to ripen their wood it may not be able to withstand the cold of winter. If they are not given a chance they cannot form good strong buds for the coming spring. They should be allowed all the time that nature has allotted them in which to store up plant food for the bursting of the buds and the growing of the new leaves the following spring. If they have been deprived of their leaves before this work was completed we should not be surprised if the next year's growth is weak and slow. As fruit growers and nurserymen we should remember that it is just as much the sign of growth in a tree to see the leaves fall when their work is completed as it is to see the buds burst in the spring. The leaves should remain on the tree until the growth between the base of the bud and end of the stem,—that corky-like growth,—thickens and thickens, finally pushing the leaf off, leaving the scar sealed over as it were, so that the moisture will not escape. How different the condition of the twigs after the nurseryman has finished stripping them. At the base of each bud is a wound made by tearing away the leaf stem before nature's time, from each of these wounds moisture having escaped the twigs have become wrinkled and dry. As we have frequently said, the drier a tree becomes the nearer dead it is.

TREATMENT OF DRIED OR SHRIVELLED TREES.
—When trees are received in this condition they should be buried completely in loose, moist soil, and allowed to remain until they regain their plumpness. They should never be put in water and left to soak for twenty-four to forty-eight hours as is sometimes practiced, with bad results, as the soaking causes the bark to separate from the wood. Should they be received in packages and frozen they should be allowed to thaw out in the package. A good plan is to place in a cellar.

HEELING IN.—When trees are received from the nursery they should be heeled in at once, even if planting is expected to begin the following day. If they are to remain over winter a trench should be opened east and west, drawing or throwing the dirt out to the south side. It should be deep enough to allow the roots to be slightly beneath the natural surface. Then the trees should be laid in with their tops to the south, allowing almost or quite the entire length of their bodies to rest upon the soil. Remove all wrappings, as straw or moss will make an inviting harbor for the mice, and they can soon ruin the trees by eating the bark of both roots and bodies. Spread the bunches, for if left tied up it will be hard to fill all the spaces in the center of the bunches with dirt, and if any should be left open the roots will dry out and be damaged in that way as well as by freezing. Place stakes between varieties so as to keep them separated.

LABELING.—The wooden labels which came on them will probably be gone before spring. So it may be best to re-label them by taking an old tin can, cutting it in strips, one end of which should run to a point, then with a scratch-awl or point of a pen knife write the name of variety on bright side of the tin, coil the slender end of the strip around the twig, the next spring (after the rust has gotten in its work) the writing will be as legible as though it had just been written with pen or pencil. These labels will be found very useful in orchard, vineyard or nursery. Scraps of zinc may be used when at hand and written on with common lead pencil which remains for a long period. After the trees have been placed in the trench, draw the dirt up from the north side well over the roots, and the greater portion of the tops, being careful to fill all openings among the roots so as to exclude the air. The placing of the tops toward the south and drawing the dirt up from the north side forms a slight mound sloping toward the north; this is of special advantage as a means of protection against the alternate freezing and thawing that often occurs during the cold nights and warm days of winter. This north slope aids in keeping the trees nearer one temperature than would be possible if the mound was reversed. Trees are often very much damaged by not being properly heeled in during winter.

If mice should become troublesome, place plank eight or ten inches wide around the trees, allowing them to lean out at the top, the edges being set in the ground two or three inches, the planks held in place with stakes. A pen of this kind will insure against mice.

A coating of saw dust, or stable manure that is free from straw will serve a good purpose to hold an even temperature in the soil if applied after the dirt has become frozen, as it will aid in keeping the frost in the ground.

To Prevent Growth.—Should the planting be delayed and the trees begin to show signs of growth, we have found it a good practice to simply take them up and heel in again. This moving will break their hold on the soil and thus check or prevent their starting.

It often happens that weather, soil or other conditions cause the planting to be delayed for many days, and trees frequently start to grow, sometimes even develop leaves while heeled in. This uses up much of the stored plant food that will be needed after the trees are put in their permanent places, as these leaves are sure to perish in the transplanting and it will cause a great effort on the part of the tree to produce others. A frequent moving may prevent great damage and often the entire loss of many trees in late planting.

THE PLANTING OF THE APPLE-TREE.

"Come let us plant the apple-tree.
Cleave the tough greensward with the spade;
Wide let its hollow bed be made;
There gently lay the roots, and there
Sift the dark mould with kindly care,
   And press it o'er them tenderly,
As, round the sleeping infant's feet,
We softly fold the cradle-sheet;
   So plant we the apple-tree.

What plant we in this apple-tree?
Buds, which the breath of summer days
Shall lengthen into leafy sprays;
Boughs where the thrush, with crimson breast,
Shall haunt and sing and hide her nest;
   We plant, upon the sunny lea,
A shadow for the noontide hour,
A shelter from the summer shower,
   When we plant the apple-tree.

Practical Orcharding On Rough Lands.

What plant we in this apple-tree?
Fruits that shall swell in sunny June,
And redden in the August noon,
And drop, when gentle airs come by,
That fan the blue September sky,
  While children come, with cries of glee,
And seek them where the fragrant grass
Betrays their bed to those who pass,
  At the foot of the apple-tree.

The fruitage of this apple-tree
Winds and our flag of stripe and star
Shall bear to coasts that lie afar,
Where men shall wonder at the view,
And ask in what fair groves they grew;
  And sojourners beyond the sea
Shall think of childhood's careless day,
And long, long hours of summer play,
  In the shade of the apple-tree."
  —*William Cullen Bryant.*

A Fruit-Grower's Home Among the Hills
Courtesy F. H. Ballou

ORCHARDING ON THE HILLS
Courtesy R. L. Hutchinson

CHAPTER X.
PLANTING THE TREE.

"Have you any ground to spare?
Go plant a fruit tree there;
The capital you thus invest
Will bring you handsome interest."

This is a subject that is beginning to claim the attention of the public in general. Our forests are rapidly passing away, and the people begin to see the dawning of a day when timber will be scarce unless grown as other crops upon the farms. The farmers of some parts of the country realized this several years ago, and made considerable plantings of timber, especially catalpa and locust, to be used for fence posts and railroad ties, and when passing through these sections one has an opportunity to study the plan and manner of planting which has been practiced. And often the success or failure of the plantation has been largely decided by the manner in which the planting was done. The same thing may be seen in the orchards all over the country, for the method of planting will be one of the chief factors in determining the stand of trees. Not only will it affect the stand obtained but the length of life and usefulness of the orchard may be largely determined by careful attention to a few details when the trees are planted.

WHEN TO PLANT.—This is a question upon which there is a difference of opinion among successful planters; one preferring fall, while likely a neighbor practices spring planting with equal success. The matter of soil and climate does and should enter largely into the decision of this question. If the soil is open—that is, sandy or gravelly—so that there is no necessity for allowing the subsoil to freeze and thaw in order to loosen it up, and if there is no danger of water standing in the hole and if you want to economize time you may do well to utilize the good fall, or open, warm days of winter, and do your planting then. So when the warm days of spring creep in and crowd out the winter weather even before the ground is in good order for planting, the trees may begin to take hold of the soil of their new home, and be ready to grow as soon as the weather will permit. That there are some advantages in fall planting none will deny, but to our mind there are so many disadvantages that we have practiced spring planting almost exclusively.

The soil in which we have planted was such as could be much better fitted by winter work, as already described under "Preparation of the Orchard Site." Then we found that we could take care of the trees better, easier and cheaper in the trench than we could if they were set in place over the orchard site, exposed to the ravages of the field mice and rabbits, thus necessitating the wrapping of the trees the first winter. Again on this hilly land we find the orchard sites frequently very much exposed to the wind, and if the trees are set in the fall they are often shaken about a good deal during the winter as the roots have not had a chance to anchor the tree by getting a hold on the soil. We sometimes find that they have been worked about until there is a funnel-shaped space around them, and the ground might freeze while in this condition. If there should come a rain, this space may be filled with water and then when another freeze follows as frequently happens, the bark on the body of the tree is bursted, either killing the tree or making it a cripple. The progressive orchardists of the future cannot afford to have many cripples to nurse. Another reason for not setting in the fall is that the top of the most of the trees that we plant should be cut back when transplanted. We cannot afford to make these wounds—from which so much moisture will escape during the winter when there is no chance for it to heal. While if all the top is left on, the wind will toss the tree about with the result just described. Even when planting yearlings we find it necessary to cut them back in order to form the heads at the desired height. If this should be done in the fall and we have a severe winter with a great deal of cold, dry, windy weather, we should not be surprised to find that the switches have dried out until there was not enough vitality left in them to burst their buds in the spring. So they frequently send up sprouts from near, or even under the surface, from wood which has been protected. We should always remember that whenever the air is dryer than the body or twigs of our trees that the moisture will escape through their bark. Without moisture there can be no growth. So whenever a tree is allowed, from any cause to become dry, its vitality is weakened. Many replants are caused by unnecessary exposure of tender bark, or wounds during the winter.

When the trees are moved from the nursery

row, trench or cellar, as the case may be, to the field for planting, care should be taken not to expose the roots to the air or wind any more than is absolutely necessary. Always remembering that the tree is a LIVING THING, and its strength might well be measured by the amount of sap contained in it, just as we might measure the vitality of a man by the blood in his veins. We say a root is dead when there is no sap in it, or alive when green and full of moisture. We should protect the roots of plants and trees in some way, either by covering with a wet blanket or moist straw, from the time they are taken up until they are ready to be put in place in the orchard.

>"Just fancy a mythical story
> Of trees that have feelings and words
>To express their varied emotions,
> Do you think 'twould be really absurd?
>How the long promised resurrection,
> From dead winter to vigorous spring,
>Made their hearts' blood circulate freer
> And their leaves to the soft breezes fling."

ARRANGEMENT OF VARIETIES IN THE ORCHARD.—This has not been as carefully studied as it should be. Not until a mistake has been made and the orchard comes to bearing does its full importance dawn upon the planter. The varieties should not only be planted in the most congenial soil, but special attention should be paid to their arrangement as to the succession of the blossoming period. Spraying has become a science as well as a necessary art. In order to spray scientifically the work must be done at certain stages of development of bud and bloom. Hence the great necessity of arranging varieties so that the spraying may begin on one side of the orchard with the earliest blooming varieties and be completed on the latest bloomers at the opposite side of the orchard. This will prove very advantageous in the economizing of time and labor.

Care should be taken that the variety rows run with the contour of the land, or the way the spray pump will have to be driven, for if it becomes necessary to drive across the variety rows, coming first to an early and then a late bloomer, it will either make a more expensive or not as effective application. Very much will depend upon the location of varieties as to the time of blooming in regard to spraying as to its effectiveness, cheapness and convenience.

PREPARING THE TREE FOR PLANTING.—

"Each injured root 'part' cut away,
But leave the rest untouched, I pray,
Then to offset the losses here,
Prune shapely too 'the top' nor fear
One-third the twigs to sacrifice—
But spare the limbs of larger size.
Treatment like this, a few years hence,
Will bring you luscious recompense."

A 2 Year Tree Prepared for Setting

Before setting, all bruised and broken roots should be removed, remembering that the plant is better off with a few short healthy roots than with a mass of long, bruised ones. It will take more strength from the tree to attempt to rebuild these injured parts than it would to throw out new ones. All cuts that have been made by removing the tree from the nursery row should be renewed. Fresh wounds heal more readily than old ones. In renewing these or making fresh ones all cuts should be made from the under side. This allows the cut surface to come in more perfect contact with the soil, then it lessens the liability of damage by the soaking in of water at the end of the root between bark and wood and causing the root to die back as is sometimes the case when the cut surface is up. For the same reason they callus much better, and this is of great importance. In order that all the cuts may have as near the same slope as possible, place the tree under the left arm in position so as to be whittled with ease, then with a sharp, stout knife, prepare the tree for setting as just described.

DEPTH OF PLANTING.—This should be governed somewhat by the contour of the land. Trees should always be set some deeper than they grew in the nursery row, for the loose earth will settle down around them and we shall find that they are not as deep as they looked to be when planted. If the land is level they should be set from one to two inches deeper than they grew in the nursery. If the land is steep they should be set two to four inches deeper. As on the steep land the dirt will not only settle, but is liable to be washed and worked away from around the trees. We should remember that a tree never gets any deeper in the ground than it was planted. While some may be planted too deep, we believe more suffer from too shallow planting. While some may suffer from being smothered by deep planting, there is more danger of damage by severe freezing or danger from drought by extreme shallow planting. While the extreme shallow planted trees often suffer more from the effects of the woolly aphis than those whose roots are deeper in the ground.

After deciding upon the depth the tree is to stand when planted, begin setting it several inches deeper than you wish it to remain. Then while one man throws in the dirt the other who is holding the tree in position and firming the soil, after a few shovelfuls of earth are in place, should lift the tree an inch or so, at the same time giving it a gentle shake. After filling in more dirt, give it another lift and shake, and so on until the tree stands at the desired depth. In this way the fine dirt will be sifted in among the roots and all the air spaces will be filled quite as well as if the dirt had been worked in with the fingers. When this finger work is left for the average planter to do, it is seldom done; he simply fills up the hole, firms down the dirt and calls it done, frequently leaving the ends of the roots turned up, caused by the tramping of the dirt down near the body; while in the plan suggested,—the lifting of the tree several times —not only insures the complete filling of the spaces among the roots, but assures the planter that the root ends are all pointing downward, which places them in the best possible position to brace the tree.

PUDDLE BEFORE SETTING.—The practice of dipping the roots of trees in a mixture of water and clay—called puddling—before setting, is one that is of great value, especially when the soil is dry, or if the trees have begun to show signs of growth in trench, cellar or package. The puddle should be made in a vessel that may be carried easily,—a galvanized tub preferable. It is made by stirring pulverized soil into water until it is about the consistency of good, heavy lead paint. After pruning the roots, dip them in the mixture and set them at once, or the puddle will dry on them and prove a detriment rather than an advantage. The purpose of the mixture is to add moisture which the tree must have before growth can begin. The coating causes the fine earth to adhere to the roots and they are able to begin growth at once. If it should be allowed to dry before the tree is set, it will require more moisture to soften it and reach the roots than it would have taken to start a growth without the puddle.

Two or four rows may be set at once, thus saving the labor of carrying the tub through the field so often, and the trees may be taken directly from the tub to the holes.

POSITION TO SET THE TREE.—Much of the future usefulness of the orchard may depend upon the mere position in which the trees are set. This is strikingly shown in many of the old orchards all over the country. It is often not even necessary when passing through a fruit section to ask from which direction the prevailing winds come, as the leaning trees with their dead sides stand as constant reminders of the effect of the winds together with careless planting. Both of these troubles might have been obviated by placing the trees in a proper position. That is when setting lean them toward the prevailing winds. An angle of 45 degrees often proves none too much, as by the time the trees are two years set they will stand straight and by this time if proper care has been given to the formation of the top it will shade the body so that the danger of sun scald will be passed and the tree has become anchored and will hold its position.

SUN-SCALD. This trouble from which so many trees suffer is frequently brought about by winter conditions. When the body of the tree is leaning from the sun it often happens that the bark on the exposed side will thaw during the warm hours of the winter days,— which are from 12 to 2 o'clock,— while the bark on the opposite side of the tree is still frozen. When night comes on this bark is, of course, frozen; the next day this occurs again, and so finally after many such changes, the bark on the exposed side is killed, and sticks tight to

Uprooted Tree Showing Where Roots Form Braces Rather Than Anchors

the wood. When the tree starts to grow in the spring this bark cannot expand so must burst. This crack not only allows the water to enter but makes an ideal place for the work of worms,—borers,—so it often happens that we find trees, before they arrive at the age of profitable bearing, ready to break with the weight of their foliage, especially if a wind storm should strike the orchard. Even should the tree be strong enough to hold up its load we have no right to expect as good fruit from this diseased tree with about one-half of its original bark surface, as from the one with healthy bark, for the conveyance of the plant food from the root to the foliage.

We cannot be too careful in setting our trees so that this trouble may be avoided.

LONGEST ROOT AS ANCHOR.—The more equally the root system is distributed around the tree the better we should like it. But if there should be one side with longer roots than the other, we should prefer placing the tree in such a position that the longest roots will be toward the prevailing winds. Remembering that they should anchor rather than brace the tree. We see this plainly illustrated in the forest trees. Often when passing through the woods we find trees that have grown against a ledge of rocks, so of course have no roots on that side. We find these trees blown over and yet none of the roots broken, but simply peeled out. While if the wind had come from the other direction so the roots would have acted as an anchor rather than as a brace, they would have had to have broken before the tree could have blown over. So we should anchor the tree by placing the longest and best roots toward the prevailing winds rather than trying to use them as braces.

HEAVY SIDE OF TREE.—We should always prefer placing the heavy side of the tree—generally the lowest limbs—towards the prevailing winds. It sometimes happens that the longest roots and the heaviest limbs are on opposite sides of the tree. In such cases we should always give the root system the preference, as the top may be changed or controlled by pruning.

After the tree is in position and the soil placed around it and well firmed down, make a level space about it from three to four feet in diameter and fine the surface as if small seeds were to be planted. The first year's growth will depend very largely upon how well the moisture is retained. For this reason it would be well to maintain a dust mulch around the trees during the growing season. Much depends upon the results of the first season's growth. If a good stand of trees and healthy growth can be secured the first season, the chances for the success of the orchard are very much better than when there is only a half stand of trees and many of those barely alive. When we meet with such conditions we are apt to become discouraged. Whenever a man is discouraged it will not be long until his business will show it, we care not what he may be engaged in. Not only will it discourage the grower himself, but every poorly grown orchard has its effect upon the development of the business in that section. When we see a fine crop of anything, that promises to pay the grower a profit, then every one is ready to engage in the production of that crop. But when we see a failure—especially in orcharding—we are apt to attribute it to the business rather than to the man, or the way in which the work has been done. Poorly planted trees have often been the cause of retarding the development of the fruit industry in a state, county or neighborhood. When contemplating the planting of trees it should be considered from various view points. Money should not be the only object of the tree planter. He should be able to get pleasure out of the work as he goes along. It should be considered a privilege to plant a tree, something that will live on and continue to be a source of pleasure and profit to future generations after we have passed from the field of action. The planting of a tree shows faith in our Creator, for we would not plant if we did not believe He would reward diligent labor with a liberal harvest. The planting of trees and orchards show that we have faith in the section or locality which we have chosen. By our choice we show that we have faith and believe the soil and climate are congenial to the growth of the varieties selected. The planting of trees should be an indication of faith in one's self, for if a planter has not faith in himself, if he is not fully convinced that he has the ability to plant the tree carefully, cultivate constantly and gather the fruit tenderly, he had better not enter the field of tree planting. He may prove a planter, and a planter only, for all trees which are planted do not yield profitable harvests.

"What does he plant who plants a tree?
　He plants a friend of sun and sky;
He plants the flag of breezes free;
　The shaft of beauty towering high;
He plants a home to heaven anigh
For song and mother-croon of bird,
In hushed and happy twilight heard—
The treble of heaven's harmony—
These things he plants who plants a tree.

What does he plant who plants a tree?
　　He plants cool shade and tender rain,
And seed and bud of days to be,
　　And years that fade and flush again;
　　He plants the glory of the plain;
　　He plants the forest's heritage;
　　The harvest of the coming age;
The joy that unborn eyes shall see—
These things he plants who plants a tree.

What does he plant who plants a tree?
　　He plants, in sap and leaves and wood,
In love of home and loyalty,
　　And far-cast thought of civil good—
　　His blessing on the neighborhood
　　Who in the hollow of his hand
　　Holds all the growth of all our land—
A nation's growth from sea to sea
Stirs in his heart who plants a tree."

A Profitable Apple Tree
Courtesy W. E. Rumsey

A WELL CARED FOR ORCHARD, WHERE CLEAN CULTURE HAS BEEN PRACTICED
Courtesy W. E. Rumsey

CHAPTER XI.

CARE AND CULTIVATION.

"Over and over again,
No matter which way I turn,
I always find in the Book of Life
Some Lesson I have to learn."

This is a subject the importance of which the beginner can scarcely realize. It is also difficult for the orchardist of experience to even intimate its importance. There are so many varied conditions which will come up for consideration, things upon which so much depends that after all that has been read and written and all that years of experience may have impressed on the grower, even then, all that we may say will only be offered as a few kindly suggestions. Hoping that they may cause some to think for themselves and thus help them to solve their own problems, for as has been so well said, "Thoughts are forces, living messengers of power."

Care and Cultivation are so closely allied that they may be well considered together; not only are they closely connected, but we shall find by careful study that one without the other will not bring about satisfactory results. We might well say they are closely connected, for when we are caring for our orchards we are frequently cultivating them, and when we are cultivating we are certainly caring for them. So when we speak of care and cultivation of the orchard there are many things that should suggest themselves to us. Such as the stirring of the soil for various purposes, namely, the destroying of weeds, retention of moisture and setting free the plant food which is in the soil. Then we must supply plant food, either by means of commercial fertilizers or barnyard manure.

The care may suggest the maintenance of fertility by the growing of some of the many cover crops, either winter or summer. As we think of these crops, the long list of legumes such as peas, beans, clovers, vetches, etc., seem to pass before us.

Then erosion is suggested, for who has not

had his lands to suffer the almost irreparable loss of soil by washing. This suggests to us at once the growing of plants as soil binders, plants which will keep the soil filled with living roots which will be active even during the winter. Plants which will be able to grow with the slightest warmth of a mild winter day; plants whose roots will be able to take up any food that is set free by the action of the elements and store it in their tissues, which after they decay, will be ready for the use of crops that may come after them. Crimson clover, with its long deep roots covered with nitrogen-bearing nodules; and rye with its rootlets forming a sod that prevents washing and at the same time making a growth that will aid future plants by its decay.

All these and many other things pass before us and many are the pictures in our mind's eye of the different kinds of care demanded by the orchard, of men taking out borers, wrapping the trees to protect against rabbits, etc., etc. Then we see the painstaking orchardist thoughtfully and carefully removing such limbs as interfere with the proper development of the tree. In another picture we see an orchard covered or mulched with a blue grass sod. Or perhaps one in which the moisture is conserved by a dust mulch maintained by constant cultivation. Every experienced fruit grower when allowing his mind to revert to the care and cultivation of his orchard, will have before him a panorama of fast moving pictures of the days and years that have passed and gone. He will be able to see those trees as they looked when set, then as they grew they seemed to carry him with them on and on year after year till the harvest.

These pictures will vary according to the section in which the orchard was grown. If it were in the great irrigated section of the west, the grower might fancy himself among the ditches and even hear again in his imagination the running of the life-giving water in the flumes. If in the central west, he may turn from the scenes of his orchard and gaze upon a great expanse of prairie rolling away in the distance. But if his lot were cast on rough or rolling lands, the picture as he recalls the landscape will be varied indeed. Here it may be a wild and unbroken forest, yonder a deep ravine, with its sparkling brook and its laughing waterfalls, or he may turn from the orchard and his eyes rest as of yore on the distant smoke-covered mountains.

Wherever these pictures are painted, whether

it be in our mind's eye of the future, or whether they hang on memory's walls, we must all realize that every section and each orchard presented then, and presents today, its own peculiar problems as to care and cultivation. And every grower must solve these problems according to the rules laid down by climate, location of site, contour of land, the nature of the soil, and the varieties of the fruits to be grown. For in this,—the Care and Cultivation of the Orchard,—as in many other things that we are called upon to perform, there cannot be any iron clad rules. There are, however, some underlying principles which might well be considered by way of suggestion.

The question is frequently asked when should the care and cultivation of the orchard begin and how long should it be continued? The care of an orchard should begin with the selection and planting of the trees and continue together with the cultivation, as long as the orchard is expected to be profitable. It would be just as reasonable to plant corn and allow it to take care of itself and expect it to return a handsome profit at gathering time, as to plant an orchard and expect it to be a profitable investment without care or cultivation. We should not be surprised that whenever we stop caring for our orchards they cease to be profitable investments. The sooner those who plant trees realize this the better it will be, not only for them but for the business.

Not only should the cultivation begin early in the life of the tree, but it should be begun as early in the season as possible, remembering that it is in the early growing months that we want our trees to make their wood growth. The earlier the ground is stirred the sooner as a rule will this growth begin. We often cultivate to aereate or warm the soil, as when we practice making ridges or mounds of earth and allowing them to remain several days then levelling them down, thus making hills in which to plant melon or ridges for other seeds.

WHY WE CULTIVATE.—One purpose of cultivation should be to improve the physical condition of the soil. So as to extend the feeding area of the roots of our plants, realizing that the finer the particles of soil the better the air can circulate through it, consequently the better growth the plants are able to make. All must realize that without soil air, plants cannot thrive, as is so plainly shown in water-soaked soils. If we would allow ourselves to think of such soils as air tight and the plant roots sealed up we would come nearer realizing the importance of soil areation than many of us do.

CULTIVATE TO SET FREE PLANT FOOD.—Cultivation may be used as a means of setting free the plant food that is in the soil, as has been said, "tillage is manure." While this is not literally true, we must all acknowledge that proper and constant tillage will increase our crops. In fact proper tillage may double the yield of crop as compared with that which received little or no cultivation, even if all the land were equal in fertility.

CULTIVATE TO DEEPEN SOIL.—The importance of deepening the soil by means of cultivation seems to be seldom thought of by farmers and fruit growers. They too often plough the same depth year after year, and not infrequently by so doing they form a hard bottom or place of separation between the turned furrow and the subsoil, which in some cases is very detrimental to the growing crops, as it may cause the water to percolate much more slowly than it would otherwise have done, being held about the plant roots too long. Again the moisture will not be brought up as readily or freely by capillarity as if the hard or glazed surface was broken up. In one sense we can increase our acreage by increasing the depth of the land which we cultivate. The day has come when the successful agriculturist must intensify his work, whether he be a grower of corn or apples. Yes, we must deepen rather than broaden our acres, get more from less land.

CULTIVATE TO INCREASE MOISTURE HOLDING CAPACITY.—We should not only cultivate with a view of deepening the soil, but we should have in mind that important feature of cultivation to increase its moisture holding capacity as well. In order to do this we must fine the soil, for the finer the particles of soil the more water it will hold. The more particles there are the more moisture it will take to wrap them. We should think of soil moisture as a film surrounding each particle of soil, for after all, no matter how deep and good a soil may be, unless there is moisture present plants cannot take their food, or as we say, they cannot live. The arid regions are striking examples of this, with their deep, loose soils, they are but deserts until the water is applied.

"I heard a farmer talk one day,
  Telling his listeners how
In the wide, new country, far away,
  The rainfall follows the plow.
'As fast as they break it up, you see,
  And turn the heart to the sun,
As they open the furrows deep and free
  And the tillage is begun,
The earth grows mellow, and more and more
  It holds and sends to the sky

> A moisture it never had before,
>   When its face was hard and dry.
> And so wherever the plowshares run,
>   The clouds run overhead;
> And the soil that works and lets in the sun,
>   With water is always fed.'
> I wonder if ever that farmer knew
>   The half of his simple word,
> Or guessed the message that, heavenly true,
>   Within it was hidden and heard?
> It fell on my ears by chance that day,
>   But the gladness lingers now,
> To think it is always God's dear way
>   That the rainfall follows the plow."

CULTIVATE TO RETAIN MOISTURE.—When we cultivate to retain moisture we should have in mind how the moisture escapes, and consequently what is necessary to prevent its escape.

We should remember that when it rains and the fine particles of soil are dissolved they run together making a smooth surface, a crust as we call it. As soon as the soil begins to dry, we notice little cracks, through these cracks the moisture will escape, for it is being brought up by those little capillary tubes and set free through these openings or cracks in the crust of the soil, to be carried away by the wind. So in order to prevent this it is necessary not only to break up the crust, but to shorten the capillary tubes, so when the moisture is brought up by them it is set free in this loose soil or dust and held for the use of the plant. This can best be done by constant cultivation. We call this method of cultivation a Dust Mulch. Where can we turn for a more striking example than to the west, or irrigated sections, for lessons in cultivation to retain moisture? If these fruit growers who have it in their power to apply water whenever their crops need it, find it profitable to keep a dust mulch by constant cultivation in order to get best results, how much more considerate of the retention of moisture should those growers be who have to depend entirely upon the annual rainfall for their supply. How seldom it is that an orchardist on the rough, rocky, dry, steep hill sides ever seems to give this a passing thought in connection with the methods of cultivation being practiced.

CULTIVATE TO HASTEN THE DECOMPOSITION OF PLANTS.—This is one purpose of tillage that is frequently overlooked. When crops are grown for the improvement of soil and only half ploughed down or buried (ploughed under seems to express it better, as it carries with it the idea of burying the material)—and that is what must happen if we can hope for the best results in the shortest time. This being the case the matter of cultivation as a means of hastening decomposition should receive the careful attention of agriculturists generally. Remembering that all the plant food which is locked up in that dry material, whether it be straw, pea vines, stubble or weeds, that plant food is utterly worthless until liberated by decomposition. In order to hasten this process we should mix or cover this material with soil and this is best done by some method of cultivation. In orchards where plowing is not practical the disk harrow may serve the purpose well.

CULTIVATE TO DESTROY PLANTS.—Cultivation is used as the principal means of destroying those plants which interfere with any special crop that the planter may be desirous of growing. Such plants are called weeds, not necessarily because of what they are, but because they are growing where they are not wanted. "A weed is a plant out of place." Too frequently the principal reason for cultivation is to destroy weeds. While we believe that cultivation should be the means of their destruction, we do not believe we should make the destruction of weeds the principal reason for cultivation, but rather try to follow so closely the other reasons for cultivation that there will be but few weeds to destroy.

CULTIVATE TO DEVELOP PLANTS.—The pur-

One of the Best Orchard Tools With Which to Cultivate Stony Land
Courtesy F. E. Brooks

pose of cultivation may be to develop a plant, or a portion of a plant. The operation may vary accordingly. For instance, we cultivate for leaf growth in the tobacco plant, and all efforts are directed toward the growing of the plant in such a way as to develop the leaf. In the rhubarb and celery we want the leaf stem, asparagus the leaf stalk; while in our pot plants we want blossoms, so we bend our energies in that direction. But in the care of the orchard it is necessary not only to grow the plants strong and healthy with their leaves, stems and blossoms, but the final purpose of all care and cultivation is the ripened fruit. In short, we care for and cultivate in order that our plants may be able to make the greatest development along the particular lines for which they are being grown.

GROWING PLANTS.—In the care and cultivation of the orchard we must consider the important work of growing plants from various standpoints. For as has already been said, the purposes for which they are grown vary widely. Some times plants are grown in order to protect or develop others. This is done when we grow a cover crop for the purpose of shading the ground from the scorching sun, or that they may serve as soil binders by keeping the soil in place and preventing those unsightly, inconvenient and ruinous gullies. Again we grow plants of the leguminous family: peas, beans, vetches, clovers, etc., in order that they may trap nitrogen (that most expensive element of plant food) from the soil air and store it in nodules on their roots for the use of future crops. While their stems are of great importance in both protecting and maintaining the soil.

GROWING PLANTS TO SUPPLY HUMUS.—The planter who has the welfare of his trees at heart will find it necessary to keep up the supply of humus in the soil. Continuous cultivation without the addition of organic matter, humus, (decayed vegetation) will soon render the land unproductive. Examples of this may be seen in some of our older peach growing sections where the dust mulch system was practiced without the growing of either summer or winter cover crops. This is more pronounced on steep, rough lands where the humus has been washed as well as worked out. We often lose sight of the importance of shading our land by some growing crop, for we know that hot sun reduces the humus supply in our soil very fast.

Humus improves the physical condition of the soil, thus aiding very materially in increasing its ability to retain moisture. For as has been said before, nature shows us how important it is to keep our lands covered with some growing plants. How quickly the old bare field covers itself. It may be with some plant that we do not think very much of, but it will serve as a cover and add some humus. The more humus a soil contains the greater per cent of moisture will it hold as compared with its weight. For example, a hundred pounds of sand will not hold as much water as will a hundred pounds of leaf mold. Humus might well be thought of as a sponge to hold moisture. It also enables the plant food of the soil (that which nature has placed there) to become available. It is the key to nature's great store house of plant food, which when unlocked and properly cared for will be ample for the use of all future generations. So when thinking of growing plants for the purpose of improving the soil physically and otherwise, we could do no better than to call to mind the idea given by one of the students of the past, which was, if he were to attempt to improve and fine the soil after his years of study and experience, he would not depend upon such mechanical means as mallets and harrows, but simply grow plants.

As orchardists we should grasp this idea, remembering that it is in the forest where plants have grown unmolested for generations that we find our loosest and as we say, our richest soils. Here in the deep, dark woods, where plants have lived and died year after year and generation after generation, is taught a lesson that we might do well to learn. It is by their lives that the future growths have been made the more beautiful and useful. Plants, shrubs, and trees apparently seem more mindful of a duty and we are often persuaded that they perform it with more certainty than many tillers of the soil, whom we call farmers. For we find plants using the soil for a life time and leaving it in a better condition for future generations than it was when taken possession of. Such is the work of the unmolested growing plant, and we as orchardists should not be less mindful of the duty we owe to the coming generations.

Then let us grow plants in order that other plants may flourish, blossom and fruit, that other men may, as we have, enjoy a bountiful harvest, which feeds and clothes our families and gives to the world her stores of grain and golden fruit.

"Oh, the good old yeller apple,
What a friend you are to all;
How we pull you from the branches
In the cool and snappin' fall.

> How we store you in the cellar,
>   Where you lay till winter time,
> When we bring you from yer hidin'
>   An' we eat you in yer prime.
>
> Yes, them long old winter evenin's,
>   How a feller's heart will yearn
> For the time that's allus welcome,
>   When the old log fires burn,
>
> An' the family gets together
>   When the weather's chill and cold;
> Oh, the good old yeller apple
>   Is as good as yeller gold."
>   —Nellie Russell Ferguson.

GROWING PLANTS TO CONTROL THE GROWTH OF OTHER PLANTS.—It sometimes happens that well tilled and cared for trees fail to produce satisfactory results. The orchardists may continue to bestow his labors but to no avail. We frequently see trees which have received constant care and cultivation during the entire summer come out of the winter in bad condition, with bursted bark and dead or dying limbs.

Again we sometimes find orchards which have long since arrived at the age of bearing and yet have not produced fruit. The location is studied and finally the conclusion is reached (frequently without being able to tell why), that the rich, deep, moist soil is causing too rapid a growth. Let us look at the orchard.

How loose and moist the soil, the leaves of the apple trees are large and thick, and are still a very rich, dark green; although the maple and elm have put on their coats of many colors. The apple twigs are slender and each soft, tender end carries young, immature leaves, while those of the elm and maple if examined would be found to be thick and woody, each terminating with a well-matured terminal bud. The difference in their condition in the spring will be that the apple will be more or less frozen back or winter killed, while the elm and maple will be unharmed. Because the wood of the apple was immature and full of sap when the freezes of winter came on, while the wood of the forest trees had ripened or matured and was able to withstand the winter.

Often such troubles are attributed to too rapid growth in young orchards and the grower is advised to crop the ground heavily and by removing annual crops lessen the fertility of the land, thereby checking the growth of the trees. It should be remembered that it is very rarely, if ever, that trees make too rapid a growth, but it is a common occurrence for well-tilled, fertile lands to encourage too long a growing season in young trees, not allowing them to cease their growth early enough in the fall to ripen their wood and form strong, vigorous buds ready for the work of the coming season. Consequently they go into the winter with half-grown green leaves on their tips, only to be frozen, in leaf and twig so that the pruning shears will have to be used vigorously in the spring, to remove the injured wood with its brown pith. Frequently the bodies are so green and full of sap that the freezing bursts their bark and the tree is injured or killed.

Thus we conclude it is because of too long, rather than too rapid a growth, that these injuries occur. We should mark well the difference between the too rapid and too long a growth. The remedy suggested would be to cease cultivation earlier in the season, if this did not prove effective and the weather continued very seasonable, (that is, an abundance of moisture) then it might be necessary to resort to the practice under discussion, namely, the growing of plants to control the growth of other plants. In this case the object should be to utilize the surplus moisture by seeding to some quick growing plants such as oats. These oat plants will use the moisture and thereby deprive or rob the trees to such an extent that the growth will cease, the wood will ripen and the buds mature ready for winter. While before, wood growth was encouraged at the expense of everything else.

Then when we sow crops in our young growing orchards we should bear in mind these two lessons. First, the growing of plants in order that others may grow; second, the growing of plants to check the growth of others. Then there is the older orchard which has been cultivated for so many years, and all to no avail. It presents another side of the question, for its leaves ripen before freezing weather and its wood does not winter kill. These conditions would seem to be ideal for the production and ripening of wood, but we must go a step further, for we do not only want wood, but the purpose of orcharding is to grow fruit. In order to have the fruit, the tree must produce flowers. Before the flowers, must come the bloom buds, just as surely as the leaf bud precedes the leaves. Then in this case, as in the other, let us see what these trees are doing. We found that the others were producing wood at the expense of fruit buds. It may be the same thing is true in this case. If so, we shall know what to do, but if on the other hand we find that the trees are producing blossoms every year, if they have been loaded with flowers spring after spring, and still no fruit, the fault lies at the door of the orchardist. When a tree has developed leaf, wood and flower buds, it has done its part in the attempt to reproduce itself, and it often happens in this day of fungous and insect troubles that unless the orchardist is considerate enough to furnish some protection, these delicate flowers will fade and fall, leaving the tree bare of fruit. This matter of protection of flowers will be treated more fully under the subject of Spraying.

However, if the trees have failed to blossom, but having made and ripened the wood growth at the expense of fruit production, it will become necessary that the grower should change the course of its energies and direct them towards fruit production by discouraging the wood growth. Remembering that there cannot be any formation of flower buds until wood growth ceases. As long as switches continue to increase their length there will be no flower buds formed, so it follows that the growth must cease or be checked. There are many ways in which this may be done. Anything we do to lessen the supply of plant food for the tree will have a tendency to cause it to form flower buds. Whether it be the cessation of cultivation, or the growing of some moisture robbing plant, as has been already described in the case of the oats. It might be done by reducing the leaf surface by summer pruning, and thereby lessening the tree's capacity for preparing plant food. Whichever plan may be resorted to, it should be kept clearly in mind that in order to succeed in fruit production, wood growth must cease early, so that the leaves will have time to not only elaborate plant food enough to form the flower and leaf buds, but they should have a chance to store up enough plant food in root, trunk, bough and bud to enable the tree to make a good strong start in the spring.

No growth can come from the plant food that is in the soil until the new leaves are formed. This early growth must be insured by the work of the leaves the season before. For this reason we place great stress on the health and protection of the foliage as mentioned in the chapter on Spraying. For the same reason the leaves should not be stripped from the young nursery trees (in order to hasten delivery) and deprive them of their development and thereby causing them to make feeble growth, or in many instances fail to grow at all.

TERMINAL BUD.—The question often arises as to how we shall know whether or not our trees are growing too late in the season. This can best be determined by the time of the formation of the terminal buds. It is not until after this bud is formed that we may have any assurance of the cessation of wood growth. When the length of the season's growth is completed it ceases to throw out new leaves, and is tipped with a large, plump bud, we are sure the tree has completed its efforts in the direction of wood production. These switches will not lengthen any more that season under normal conditions. The energy of the tree should be used in ripening wood and strengthening buds.

The terminal bud should form early enough in the season to insure the thorough ripening of the wood and full maturity of buds before freezing weather. Untimely cultivation may cause a second growth. For instance, where potatoes are grown in the orchard (and they are often one of the best hoed crops) it sometimes happens that digging the crop of potatoes gives the ground such a thorough working that when followed by seasonable showers the trees start a second growth, which result in damage to the young growth and frequently to the trees. To prevent this it is well to sow the ground, (after digging or stirring,) heavy with some crop.

As recommended under the head of growing plants to check the growth of others. Crimson clover may answer if the case is not too severe. Rye is sometimes used, but as it does not draw very heavy on moisture in its early growth, it is rather slow to correct the trouble. If used, however, care should be taken not to allow it to stand too long in the spring, as this would result in great damage to the trees. Oats may be relied upon, they should be sown very thick, as it is the young plant that must correct the trouble, or it will be too late to prevent the damage. The winter will kill the oats and the plants will add humus to the soil, which will aid in the growth of other plants. Besides the dead oat plants assist very much in holding the snow, especially in windy locations, by giving a covering to the ground that would have otherwise been bare, and in this way frequently preventing washing.

DIFFERENT METHODS NECESSARY IN DIFFERENT SECTIONS.—Not only do we find it necessary to practice different methods of care and cultivation in different sections of the country, that is, the same methods which bring profitable results on the more level lands of Illinois might be the means of ruining the lands of more hilly sections by washing. Not only do

Practical Orcharding On Rough Lands. 157

the conditions change with the sections, but we may be able to find marked differences in the nature of the needs, care and cultivation of the orchards in the bounds of a single state or even county.

CARE OF THE ORCHARD.—It has already been said that care and cultivation should begin when the tree is planted. We have had something to say of the purpose of cultivation. Now the nature of the care shall claim our attention. After the tree is in place, (see chapter on Planting) and the hole well filled, the surface should be levelled down. If on steep land it may be well to leave the upper side a trifle lower so as to catch as much water as possible. The surface should be fined thoroughly for a distance of three or four feet about the tree. This space should be stirred frequently during the first season, once a week would be a safe rule. The purpose of this cultivation is to retain the moisture as described previously. Again let us remember that whenever the soil becomes dry it cracks, and through these little cracks the moisture is constanty escaping. Without moisture there can be no growth, so this cultivation is very important the first season, as the tree has just been removed from the nursery row where it has had intense cultiva-

A Cultivated Orchard With Hoed Crop

tion. To neglect it at this time after changing its environment means a stunted growth. We find any special care given this first season counts for a great deal in after years.

CROP FOR FIRST YEAR.—The trees having grown in a more or less crowded nursery row for one or two years, as the case may be, we have found that a certain amount of shade the first year is rather beneficial than otherwise. Corn has been found to fill these requirements and is used quite generally for the first season, where the strength of the land will permit. In planting, care should be taken to leave out a hill on each side of the trees, so as not to crowd them too much. A very good practice is not to plant any corn in the tree rows at all. Sometimes the corn rows are laid off without any regard to the trees, often leaving some of the trees in the middle where they will be barked and bruised by the single trees, or by the plow, in the cultivation of the corn. The trees should always stand in the row for protection, if for no other reason. When young trees are bruised or barked they may not appear to be injured very much, but examine them years after and although the wound may be covered you will find the dead tissues are still there; a tree simply pushes a growth over the wound. Stand by a sawmill and watch the slabs as they are torn from the side of one of the monarchs of the forest. You will often see the charred spots that are sawed into, showing the effects of forest fires which have visited the woods when this, (now an immense log,) was a mere sapling. When our hand is cut or bruised it heals from the bottom of the wound, and if a scar is left it will be on the outside; but the hand or finger, while carrying a scar may be as strong as ever, but the tree simply covers over the dead wood and is never as strong as if it were all green and active. How often we find trees breaking over in our orchards with probably their first heavy crop, and on examination we find the cause to be from injury which occurred in the early years of its life. We cannot be too careful of the bodies of these trees, for it is upon them that we expect to build. They are the foundation of our future orchard, and if the foundation be weak so will the whole structure be weak. Careful cultivation of the corn crop will not be enough to insure the best results in the growth of the tree.

So we find it pays to keep the space (which has already been spoken of) about the newly planted trees frequently cultivated by the use of a steel rake. A few licks about each tree will break up the crust and conserve the moisture. This continued until midsummer has given very satisfactory results.

CARE OF TREES THE FIRST SEASON.—After the trees are in place they should receive their first pruning, which is described in Chapter on "Pruning for the Welfare of the Plant." An application of soft soap will prove advantageous in keeping the bark bright, clean and healthy, as well as elastic so when the trunk expands with the rapid growth it will not burst as we see frequently on neglected trees that have been brought into sudden growth by cultivation or fertilization.

> "Each spring take your soft soap pot
> And paint each tree throughout the lot,
> This will insure a healthy kind,
> Kill worms, lice, bugs of every kind."

In the fall the trees should be prepared for winter by taking the dirt away from around them down to the roots and examining them for the borers. If any be found they should be removed with a sharp knife, sometimes a wire is used, simply prodding the worm to death and leaving the channel covered with bark. This practice has one serious disadvantage; that of allowing the water to enter, filling the channel, which when frozen may burst the bark and cause serious damage. After locating the borer by scraping the dirt lightly from the bark of the tree with the back of a knife blade, watch for discolored spots. When one is noticed a slight pressure with the point of the blade will satisfy the operator whether he (the worm) is there. Follow the channel until the borer is removed, after which clean out the castings and pare off the edges smoothly. This will heal over more perfectly than if the channel is left filled with castings.

While the dirt is away from around the tree is the time to put on the wrappers for protection against the mice and rabbits. They should be allowed to extend well down into the ground, as the mice sometimes work a little beneath the surface. After the wrappers are in place the dirt should be drawn well up around them.

THE KIND OF WRAPPERS TO USE.—There are various kinds of wrappers on the market, such as wire window screening cut in strips the desired width. Coiled wire is sometimes used. It is a better protection against rabbits than mice, as the mice can get through it to some extent. The wooden veneer wrapper is good, and may be obtained from a basket factory. Corn stalks are frequently used, especially when corn has been grown in the orchard that season. Tar paper should not be used, as the tar is injurious to the bark. Wrappers should be fastened on securely with a small wire. If the screening is used the wire may be passed through the meshes, thus preventing its dropping down, or may be caught by the ends of their own raw edges. In the case of corn stalks the wire will have to be drawn quite tightly, or when the stalks shrink, as they will, it may drop down, and the stalks fall off just when most necessary. The wooden veneer wrappers may be threaded on one edge with a wire, thus keeping the wire in place and preventing them from being blown off by severe wind storms. The veneer should be thoroughly dampened several hours before using. This will prevent them from splitting, as well as curl them ready to be put in place.

WINTER CARE OF THE ORCHARD.—We all recognize the importance of the care of the young trees the first few winters. But many of us seem to forget that the old bearing orchard should have some winter care. This should consist of cleaning up all the cull fruit which may have been left at picking time. Also the removal of the mummies left from Brown-rot of peach, and Bitter or other Rot of the apple,

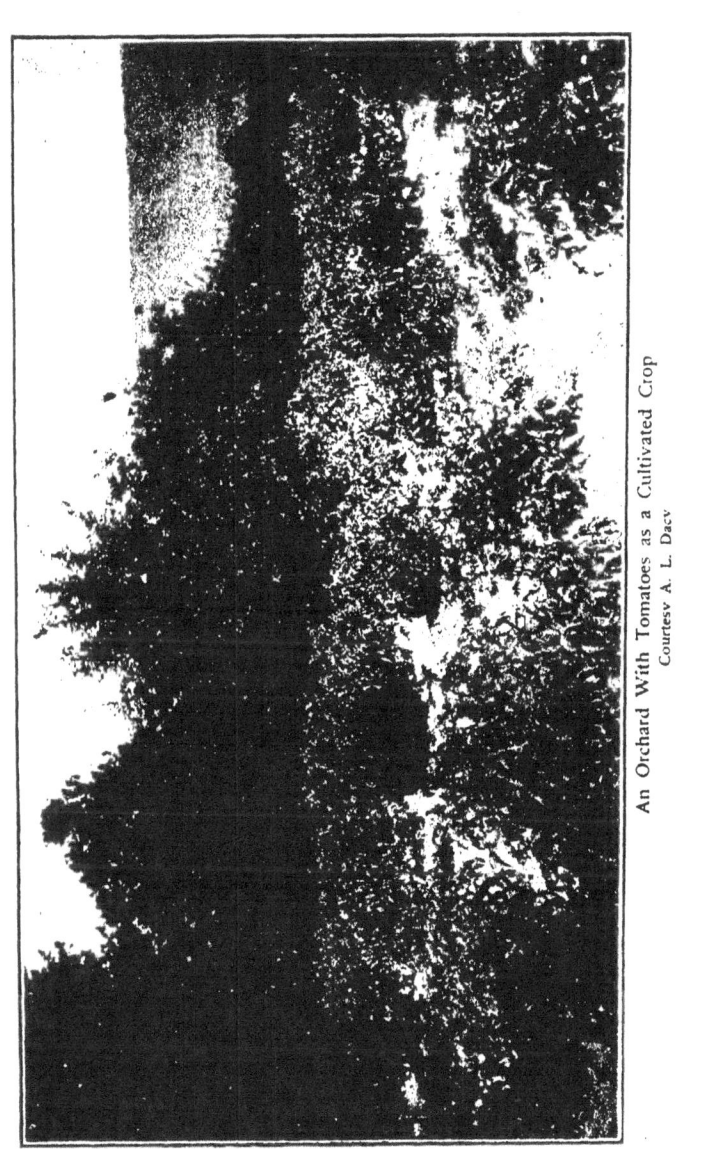

An Orchard With Tomatoes as a Cultivated Crop
Courtesy A. L. Dacy

that may be hanging on the tree or lodged in their forks. If left, these will be a source of danger to the next crop, and may make it necessary to do much more spraying than if they were not present. They should be taken from the orchard or buried or burned. The removal of the rough bark on the body and main limbs of the trees may destroy many insects which have sheltered under it. Care should be taken not to scrape deep enough to injure the trees. We sometimes see mules, cattle or sheep in orchards during cold winter weather browsing the branches, and frequently gnawing the bark from the bodies of the trees. We should remember that these trees are living things, although they appear dead because they have lost their leaves.

> "You think I'm dead,"
> The apple tree said,
> "Because I have never a leaf to show,
> Because I stoop,
> And my branches droop,
> And the dull, gray mosses over me grow;
> But I'm alive in trunk and 'root.'
> The buds of next May
> I fold away,
> To deck the end of every shoot."

CROPS TO GROW IN THE YOUNG ORCHARD.— Besides the corn, which is one of the best crops to grow in the young orchard, there are hoed crops, such as potatoes, tomatoes, straw-

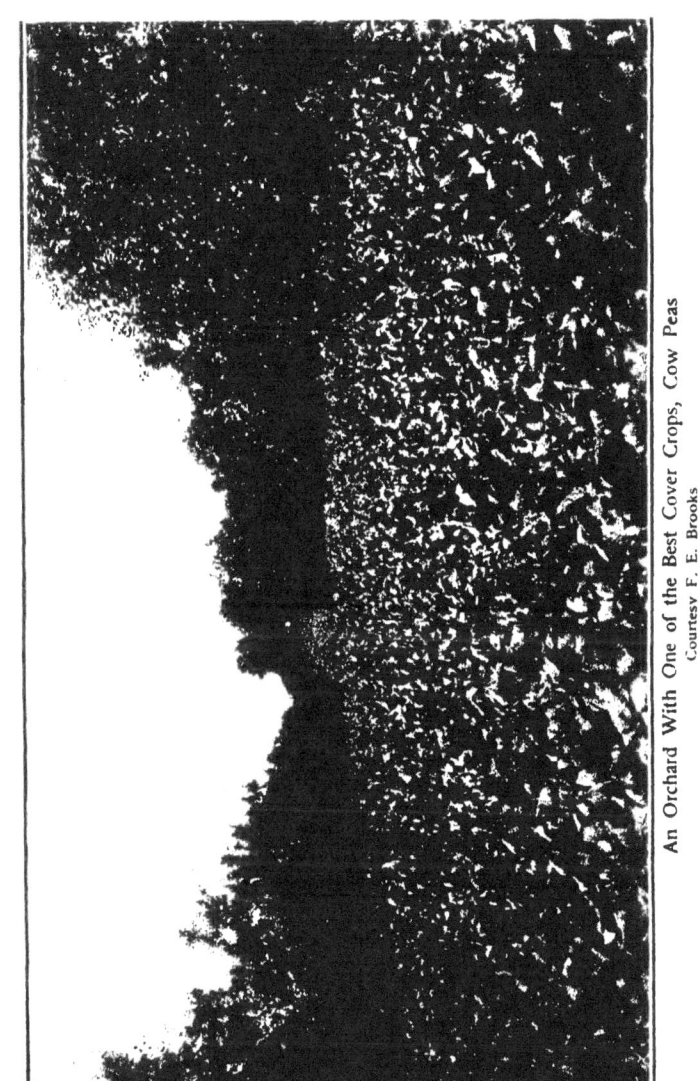

An Orchard With One of the Best Cover Crops, Cow Peas
Courtesy F. E. Brooks

berries, etc., which have met with hearty approval. Hoed crops are always advantageous to the development of the trees in their early years, as they insure good tillage. Care should be taken not to promote too late a growth by too long a season of cultivation, or a second growth may be started causing great damage as described under the head of "Growing Plants to Check the Growth of Other Plants." Blackberries and raspberries have not proven satisfactory crops to be grown in the apple orchard for various reasons, of which their habit of suckering or sprouting was by no means the least.

SUMMER COVER CROPS.—The importance of cover crops can hardly be realized until one has experienced the use of them, as has been suggested (under discussion of Humus.) Cow peas have been the most valuable, with the soy beans as close second. The clovers fill an important place. Whatever is used, be it cow peas or clover, it will be found best to leave them upon the ground in the form of mulch. Either clip and allow to lie where it falls, or it may be placed around the trees. At any rate, it should not be pastured, with even hogs or sheep, as they are liable to do the trees more injury than the pasturing is

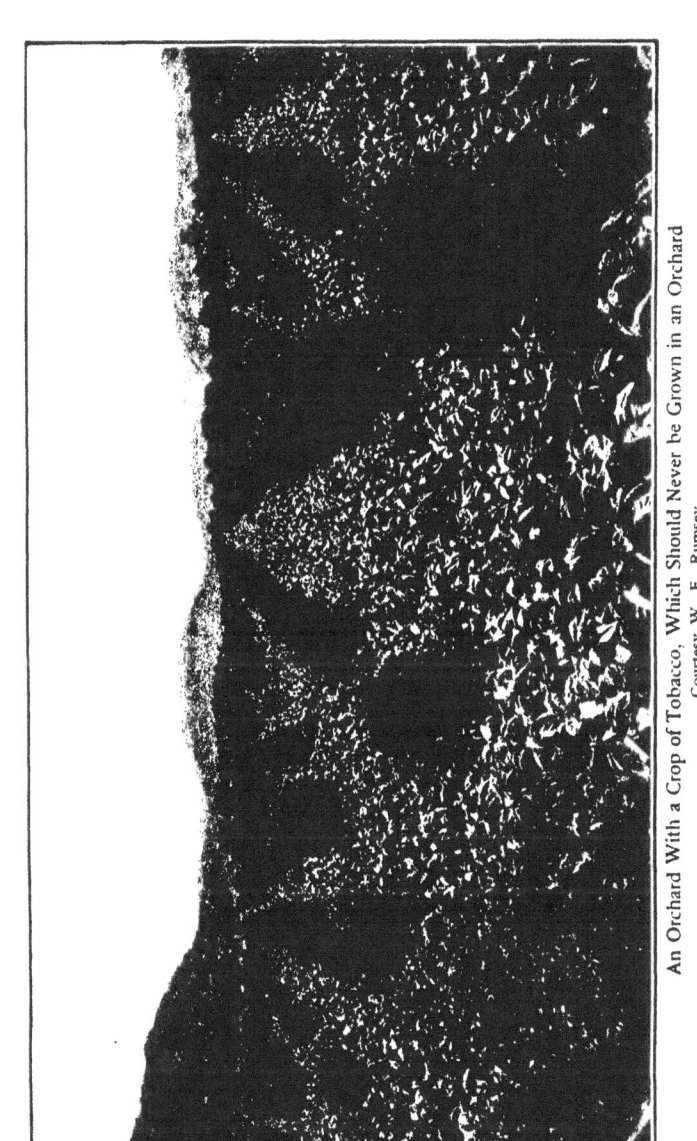

An Orchard With a Crop of Tobacco, Which Should Never be Grown in an Orchard
Courtesy W. E. Rumsey

worth, besides we can well afford to devote the entire use of the lands to our orchards. If not, the trees had better never had been set.

WINTER COVER CROPS.—Crimson clover succeeds well in some sections, especially where the falls are seasonable and the winters are not too severe. Rye is often used, but it is not good as some legumes, such as the vetches. They, however, are not practical, as the seed is too expensive. That we must use something to keep our soil covered during the winter is no longer a question, either among farmers or orchardists. Whenever possible some of the legumes should be used. But if rye should be used care should be taken not to allow it to stand too late in the spring. If allowed to make too much growth, it becomes woody and will require a much longer period in which to decay, besides it has drawn heavily on the moisture supply during the last few weeks of its growth.

TREATMENT OF COVER CROPS.—The purpose of cover crops being to aid in plant production, by adding plant food to the soil, and by shading the land and preventing washing, they should be treated as circumstances may suggest or demand. If peas are sown, either broadcast or in drills, the peas may be gathered for the next year's seed supply, and the vines left on the ground. This is especially necessary on steep land where there is danger of washing. These vines will be mashed down by the snows and make a complete protection for the soil. The cow pea being a southern plant should never be planted until the soil is thoroughly warm. Too early planting, we think, is the cause of more failures in the growing of this crop than anything else. The next spring after the weeds have started (by this we know that the soil is warm,) the vines may be worked into the soil with a cut-a-way harrow and the ground reseeded. Always leaving a space around the tree as has been described. This treatment may be continued unless the tree makes too much wood growth, then if the orchard is on steep land, some of the mulch systems may be practiced.

Winter cover crops should not be allowed to occupy the land too late in the spring, especially rye, the danger of which has been mentioned. Crimson clover and vetch should be ploughed under, or preferably cut in with a cut-a-way harrow before the growth is large enough to cause danger of souring the soil; if cut in however, there will be little or no danger of acidity, as it will not be buried so deep but that it will dry out and act more as a mulch than when turned down in heavy furrows.

MULCHES.—One of the objects of mulching is to retain moisture so we may get the best plant growth. In order to do this it is necessary to keep the surface covered to prevent evaporation. The dust mulch, which has been referred to in connection with the cultivation of the young trees the first season, is an example of this. There are other kinds of mulches that may be used to retain moisture. When thinking of mulching as done by the use of plants, either dead or growing, the object is to keep the soil dark, cool and damp, at the same time, we have in this cover, a certain amount of plant food and leave it on the surface in the form of decaying vegetation. This would bring about ideal conditions, not only for the conservation of moisture, but will enable the bacteria to perform their work of liberating plant food. We have all seen the marked effect, upon the following crop, of the shading of the ground by wheat shocks. How much looser it is and how much quicker the next crop will come up where the shocks stood, than along side where the surface has been exposed to the heat of the sun and action of the wind. The orchardist has not failed to take notice of

Gravel Land Where Clean Culture Has Been Practiced

these effects, and is taking advantage of them by the use of mulches, especially on steep, shaly hills, where the humus is well nigh exhausted, consequently the moisture is very hard to retain. Today we hear mulching frequently mentioned in connection with cultivation. So we shall take up the various kinds of mulches and consider them as they apply to orcharding on rough lands.

DUST MULCH.—There are many sections where orchardists practice clean cultivation, that is, they do not attempt to grow any cover crop, depending entirely upon tillage, not only to set free the plant food of the soil, but to conserve the moisture as well. In order to accomplish this it is necessary that the surface be kept finely pulverized or as we say, keep a dust mulch. While this has been, and is a successful method of management in level sections, it would not be practical on rolling lands, unless the nature of the soil should be particularly adapted to such treatment. There are some sections of mountains where the soil is so filled with small stone so that the water sinks as fast as it falls. Here the danger of washing would not have to be considered. But the practice of dust mulch without a cover

Apple Tree Mulched With Grass Grown in the Orchard
Courtesy F. E. Brooks

crop at some time of the year means depleted fertility sooner or later.

FOREIGN MULCHES.—It frequently happens that there is a surplus of straw, old hay, fodder or coarse manure that may be hauled and spread around young or bearing trees. This foreign mulch brought in from other fields is a very great help to the orchardist who can apply it without too much expense. There are, however, many rough, steep hills planted in orchards where it would not be practical to attempt to haul these foreign materials up the hills, if the orchardist should have them at his command. So some other means must be devised whereby the same effect may be obtained.

> "To keep your trees in copious bearing,
> Supply them food with hand unsparing;
> Rich barnyard compost, well decayed,
> Work it in well with hoe and spade
> Around the tree on every side—
> Its mouths, you know, stretch far and wide;
> Such feeding given twice a year,
> Ere long will make its worth appear."

MULCH GROWN IN ORCHARDS.—The growing of some crop between the trees is one of the practical means of obtaining material for mulching the orchard that is steep. After this is grown, it then becomes a question with the orchardist as to the way it should be used. It

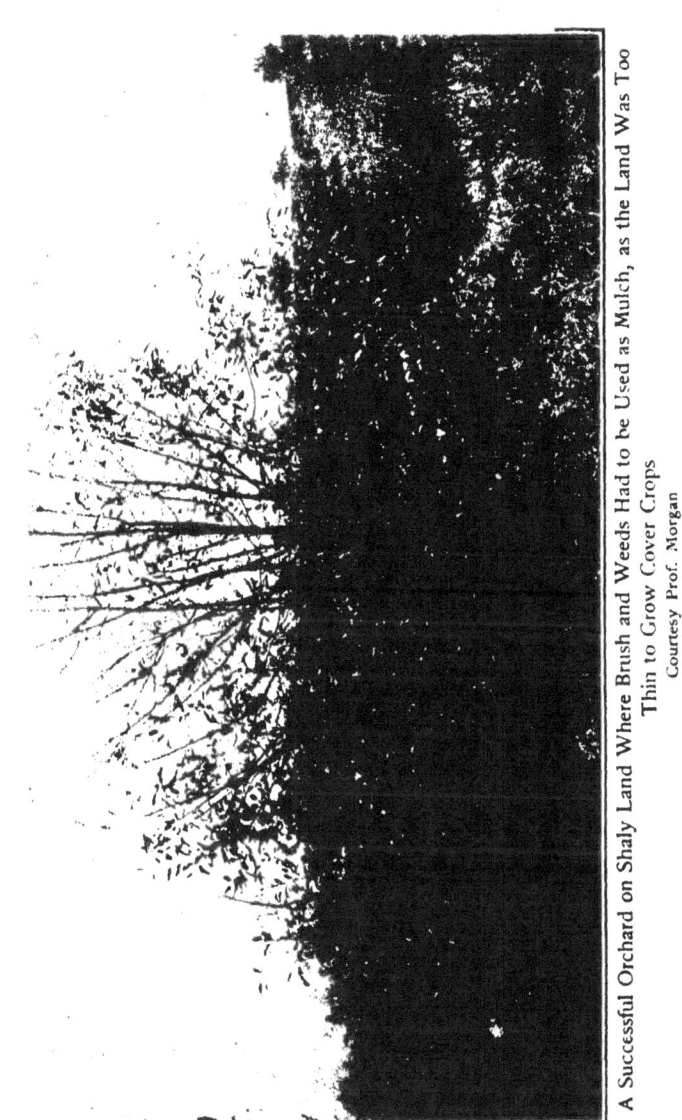

A Successful Orchard on Shaly Land Where Brush and Weeds Had to be Used as Mulch, as the Land Was Too Thin to Grow Cover Crops

Courtesy Prof. Morgan

may be cut and piled around the trees. If this is done, care should be taken not to place it too close to the tree on account of the danger of injury by mice which are apt to harbor in it. Then the feeding roots of the tree are out from the body, so for this reason as well as the one before mentioned, the mulch, whatever it may be, straw, fodder, clover or coarse manure, should be placed or spread well out under the drip of the limbs. The mulch that is to be grown in the orchard may be red clover, cow peas or even the wild growth of weeds and brush. If it is mown twice during the season and placed around the tree, each time the circle being enlarged, the roots will extend their feeding area more rapidly. The material will also decay quicker than if allowed to obtain its full growth, and form its woody fibre. This same material may be cut two or even more times during the season and left where it falls, thus making a continuous mulch over the entire surface. This method is more practical on good land. As on the thinner soils where the crop will be lighter, the shading effects of the mulch near the tree, as well as the decayed vegetation, is more necessary, and is generally placed about the trees as described.

A GROWING MULCH.—Common red clover

Orchard With Cover Crop of Red Clover
Courtesy F. E. Brooks

is frequently used as a growing mulch. In this case it will be found profitable to pasture it with a mowing machine, (that is, clip the clover frequently,) never allowing it to blossom. In this way the trees receive a great benefit the same season from the cuttings, as it forms a mulch over the entire surface and decays quickly, increasing the growth of the next cutting. If this is continued throughout the season, the clover will be found to root much deeper than when allowed to make its entire growth as when cut for hay. Besides the clover will make a stronger growth if not allowed to seed, as it like other plants, attempts to reproduce itself, and when this is accomplished it is ready to give place to the next growth of plants. Other plants may be used as growing mulches, but the common red clover has been most satisfactory.

SOD MULCH.—The subject of orcharding can hardly be mentioned now that the practice of sod mulch does not come in for its share of pros and cons. Many condemn the sod mulch system before they have given it a fair trial. That is, they plant an orchard, scatter a little grass seed around over it, then when it makes its appearance a bunch here and there, with a greater part of the ground bare and

A Sod Mulched Orchard
Courtesy F. E. Brooks

exposed to the sun and wind, the project is declared unsuccessful. Then orchardists are ready to say the sod mulch system is a failure. In order to mulch effectively the ground should be completely covered. The old fence corners in the field that have not had the grass eaten off for years, so that when it is stepped upon it is like walking through a bed of straw. These fence corners, overgrown with blue grass, are like the sod mulch orchards should be. There the grass has kept the ground dark, cool and damp. Blue grass makes an ideal permanent mulch, making its first growth early in the season, when moisture is generally most abundant. Then along through midsummer it rests, or ceases its growth for a while. Again, in late summer the second growth comes on, making an abundance of green covering to keep the fire from running through it during fall and winter. This matter of fire protection is one that should not be overlooked, especially in mulched orchards. Fire in an orchard, although it may seem little more than a smoke, often works great havoc, especially if it be about the time growth begins. Often whole boundaries of forests are completely deadened by allowing fire to run through

A Dangerous Mulch on Account of Danger of Fire
Courtesy F. E. Brooks

them about the time the leaves start, hence the importance of a living or growing mulch.

SELF MULCHED TREES.—Much depends upon the style of building which has been employed when the trees were formed as to the methods to be used in their care and cultivation. This is particularly true as regards mulching, on much of the thin, shaly soils of the steeper sections of the apple belt it might be advisable to grow the trees so that they would mulch themselves. That is, have the heads low enough that there would not be any vegetation grow under them. This has proven very satisfactory in our own orchard, which is too steep to admit of the dust mulch, or to allow the hauling in of foreign material. The land needs a continual cover. It is too thin to get a sod mulch such as would insure good results. So this self mulching plan has been practiced very satisfactorily. Mulches of different kinds are growing in favor as orchard covering, especially on steep, rolling land, and they will be used more in the future than they have been in the past.

CARE OF THE TREES.—The general care of the trees includes so much, even in a single orchard, that to attempt to give it with the variations to suit the many different locations,

A Low Headed Tree, Mulched With Its Own Limbs
Courtesy W. E. Rumsey

sites and soils, would be the height of folly. So what may be said will be only suggestive, having in mind, however, orcharding on rough lands. The first year's care of the orchard has already been given. The general care of orchard trees, especially apples as has already been said, should begin with the setting, and continue as long as the trees are expected to be profitable. This should consist of careful cultivation which has been described, as well as the care of the bodies of the trees by worming and soaping spring and fall.

MOUNDING.—Mounding, to cause the borer to locate higher on the stem or trunk of the tree is a good practice. Place a mound around the tree eight to ten inches high. This should be done in May or early June, then when the borer enters the tree at the top of the ground or mound, it will be much easier gotten out than if it were at the roots. The birds can be of even more service than if the trees had not been mounded. When the mounds are taken down in the fall and the trees examined, if there should be any worms left the birds have a good chance to take them out during the winter, which would not be the case if mounding had not been practiced. There should always be a space about the tree kept clean, never allowing any grass, weeds or mulch of any description to collect against the trunk. Protection in winter from rabbits and mice by means of wrappers as described under "First Year's Care of the Orchard," should not be overlooked, even after trees attain age. Visit the orchard frequently in winter, and if water should be found standing around the trees from any cause, it should be removed by draining, as standing water about trees frequently causes bark trouble, and that means death.

Other winter work has been mentioned under "Winter Care of the Orchard," such as removing the rough, loose bark, gathering, burying or burning the mummied fruits, etc., etc.

Pruning and spraying will be treated in following chapters.

To care for and cultivate the orchard means to supply the trees with the necessary plant food in the proper proportion for the development of leaf, wood and fruit. To do this the soil must be protected from extreme heat, drought and washing, by the growing of plants winter and summer.

Moisture must be supplied in order that plant food may be available, so that the growing of plants must be followed by a careful use of them, that they may supply the necessary organic matter to enable the soil to take hold of and retain moisture.

Cultivation should be such as will set free the plant food of the soil, at the same time endangering it as little as possible to the wasting effects of the elements.

The care of the orchard should consist of preventives, rather than cures. He who would be successful in caring for and cultivating the orchard should be determined, persistent, watchful, prompt, earnest, thoughtful and diligent.

> " 'Tis not the branches, not the leaves,
> Where tints of gold the sun inweaves,
> Nor yet the sturdy trunk, that tells
> Where life, incipient, active, dwells.
>
> Within the fructifying mold,
> That moisture from the rivers hold;
> In the warm breast of Mother Earth,
> The giant tree is given birth."
> —J. M. *Cavaness.*

A High Headed Apple Orchard Where Props Have to be Used
Courtesy L. G. Corbett

THORN TREES ON TOP OF ALLEGHENY MOUNTAINS PRUNED BY CATTLE
Courtesy F. E. Brooks

CHAPTER XII.

PRUNING.

"Watch over your trees with jealous eye
If any fault or harm you spy,
Proceed at once the cause to know,
Nor give the mischief time to grow.
Cut out each cankerous spot with care,
Until the wood shows white and fair.
Cut lengthwise to the bark, though sound,
With beeswax cover well each wound."

There are many and varied operations that the orchardist is called upon to perform in growing the trees, and in the production of a fruit crop. Each in its turn may seem to be the most important. One grower may regard this as the most vital, while another may lay greater stress on an entirely different part of the work. Had we been asked a few years ago to mention the one operation which we consid-

A Tree Built on Four Limbs, All of Which Came Out Together, One of Which Gave Way—Heart Shows Where Other Limbs Were Removed When Tree Was Planted

ered the most important, we should have said spraying, as we were firmly convinced of this. But not until after we had seen the effects of the neglect of pruning as it was shown in a finely cultivated orchard. One of the best that was to be found in one of the greatest fruit sections of the country, were we convinced that the pruning, or the building of the trees, (for that is the way we should think of the growing of the orchard,) was of more importance than spraying. In this particular orchard all the latest methods of cultivation and spraying, had been followed closely. The trees were in fine growing condition, the fruit was all that could be asked for, the crop was heavy, so heavy, in fact, that the trees were broken down, not necessarily because of the size of the crop they bore, but because of the weakness of the trees, or scaffolds that attempted to carry the load. When we saw those crotched trees with one side split down, or a tree that had been allowed to form three branches, (all starting out at one point,) for its top to be built on, and one had broken down with its load of almost perfect fruit, we were impressed with the importance of building the tree.

There, among those split and broken trees

we decided that spraying should be considered secondary, rather than first in importance in orcharding. We reasoned that this whole crop of splendid fruit might have been lost by the neglect of spraying. The Codling worm or Bitter Rot might have caused heavy losses, then with proper spraying the next season's crop could have been saved. As it was, the trees were ruined, their future usefulness largely destroyed, and no amount of labor could restore them. What a change had come over this orchard. Only a few weeks before it had been a thing of beauty, the grower was proud of the result of his labors, it had attracted the attention of the orchardists for miles around, many were the visits and compliments paid. But it was a sad sight to the true tree lover to see those trees broken or wrecked this early in life. The grower was discouraged, his hopes had been blasted, he had had dreams of a bright future, which had disappeared, as he realized his mistake in the building of his trees.

In walking through that orchard upon which so much labor had been bestowed, looking at those trees broken to pieces with their first crop, their bodies in many instances split to the ground, we resolved then and there that we would pay more attention to the building of our trees, lest we should meet with similar loss.

THE PLANTER'S IDEAL.—Whatever may be a man's undertaking, he should have an ideal. In fact, we are persuaded that without an ideal, success is rarely if ever attained. While this is true, in all kinds of business, it is particularly so in orcharding, and especially in the matter of pruning. How would the carpenter ever succeed if he should simply begin to saw off boards and nail them together without any definite plan, model, or idea in mind. He should be able, even before he drives the first nail, to see in his mind's eye the whole structure just as it is to be when the finishing touch is given. For he has had the plan so well fixed in his mind that he could tell step by step how it would look when completed.

It is the result of constant effort put forth towards the one end—his ideal. It is not enough that we have an ideal today, but we should have the same ideal tomorrow. If our efforts are guided in one direction for a time, then in another direction we cannot hope to accomplish anything, especially is this true in tree building, as it takes years to complete the structure.

If we prune our trees this year for a certain form or ideal, and then next year have an entirely different one in our mind and attempt to prune for that, how can we ever hope to

An Ideal Low, Double-Headed Apple Tree. Fifteen Years Old.

form, or complete the building of our trees. While it is important, in fact, while it is absolutely essential to success that the planter have an ideal tree in mind when he begins his work, it is equally important that he stick to it. Not only for one year, but he should keep the one ideal constantly before his mind, just as the architect does, until the building is completed. Great care should be taken in deciding upon the ideal, as it will prove much more difficult, and even more expensive, to change the plan of your trees after the building has been begun than it is to change the plans of the house, and every one knows how expensive that would be.

THE IDEAL TREE.—The purpose for which a tree is grown and the tastes or fancies of the grower, together with the variety of the tree, whether it be for ornament, or for shade, will largely determine the ideal. If for the latter, such as elm or maple, there should be enough length of stem or trunk to allow an abundance of room beneath long spreading boughs. The head or top should be dense, so that the sun could not even penetrate it. We should not fail to have our ideals even in our shade trees, neither should we fail to plant and care for them, not only for the comfort and beauty which they add to our surroundings, but for the pleasure they may give those who come after us. Planting trees is a duty we owe ourselves and our country.

"Beautiful trees by the wayside,
 Blest are their ample boughs green,
Casting their kind, cooling shadows,
 Lending a charm to each scene;
Cherish them, guard them from danger,
 Think of the good they will do;
Each one of love is a token,
 An emblem of hearts that are true.
  Beautiful trees, Beautiful trees,
  We their friends will be.

Let us plant trees by the wayside,
 Gladly our part let us do,
Make fair the pathways for others
 While we Life's journey pursue;
Each lend a hand that is helping;
 Here 'neath the broad heavens free,
Let us bestow on our country
 The gift that a blessing may be.
  Beautiful trees, Beautiful trees,
  We their friends will be.

Under the green, waving branches,
 Deep in the woodland and grove,
Hark how the birds trill their praises,
 List to their sweet songs of love.
Beautiful trees of the forest,
 Trees of the garden and field,
Spare them from needless destruction,
 That each may its fair bounty yield.
  Beautiful trees, Beautiful trees,
  We their friends will be.

Plant well the seeds, that they fail not,
 God will keep watch while we sleep;

> Soon on their freshness they'll bring us
> Joy that all freely may reap.
> Let us plant trees by the wayside,
> Plant them, with love, everywhere;
> Ours is the pleasure to give them,
> That others their blessings may share.
> Beautiful trees, Beautiful trees,
> We their friends will be."

AN IDEAL FRUIT TREE.—To our mind an ideal fruit tree would be one which could produce the largest possible amount of fruit with the least expense. Not only considering the cost of caring for the tree, but also the harvesting of the fruit. It should also be built so as to be able to hold up its load, also occupy as little space as possible. In other words, it should (if it be an ideal tree) give large returns in fruit for every foot of space it occupies. In order to do this it should have a large fruit bearing surface, and yet not be top heavy. Then it follows that it should be a tree that would produce fruit all through its centre, and not as we frequently see, just a rim of fruit over the outside of the tree.

The location in which the orchard is planted may and should govern the planter to a certain extent in deciding upon what form of tree his ideal shall be. For instance, if the plantings are to be made on level land, and it is important that there be a circulation of air under the

A Fifteen Year Old Tree With an Abundance of Bearing Wood so Distributed That the Limbs Will Not Interfere With Each Other and Will Hold Up a Heavy Load Without Breaking

heads of the trees, as it frequently is, then the planter's ideal will doubtless be a tree with a high head. If it be on level land where clean culture may be practiced without loss of soil or fertility by washing, the ideal might be (to some) a high headed tree. So the reader will readily see that location, as well as the method of cultivation expected to be practiced, may govern the grower to a certain extent in the matter of choosing his ideal. As we are dealing with the problems of orcharding as they present themselves to us in our work on rough or rolling lands, we shall attempt to form an ideal which will be of the most service under the conditions of care and cultivation of the trees, and harvesting the crop on such lands as are rough, steep, rocky and thin. Wherever the tree is to be planted, or whatever the ideal may be, the planter should be able before the tree is planted to see in his mind's eye the mature tree, ten, fifteen or twenty years old, loaded with fruit, its branches bending, (it may be to the ground), yet no broken limbs. If he cannot do this he has not gotten his ideal well enough fixed in his mind for him to be able to complete the building of the tree.

LOW HEADED TREES.—We are convinced that the low headed tree is the one that will

A Low Forked, High Headed Apple Tree

give the best results on our steep lands. When we say a low headed tree, what should be understood? Prof. L. H. Bailey puts it well when he asks the question, "One that is started low and allowed to run up, or one that is forked high and allowed to droop?" When we speak of a low headed tree we see at once a tree that has been forked low and then kept low by annual heading in until the scaffold, as it were, is formed. There is a vast difference between a low forked and a low headed tree. We sometimes see trees which have been forked low and then let grow up eight or ten feet before any bearing wood was allowed to form. This is a low forked but not a low headed tree. There are many reasons why the tree with the low head is preferable. We all recognize the advantage in spraying as discussed in the following chapter, by making it easier to reach with the spray material. Cheaper on account of less waste of material than in the case of the high heads, and the work may be done much more effectively, as the operator can see more readily when all parts have been reached. The low head is a source of economy in picking the fruit, to say nothing of the greater ease with which this labor may be accomplished. Low headed trees resist the wind much better than

A Badly Built Eight-Year Old Tree Which is Liable to Split With Its First Load, Although it Does Not Carry Enough Bearing Wood to be Profitable.

those with long stems or trunks. When the head is low and the crop is heavy, the limbs will often rest upon the ground, and this serves to rather brace the tree in time of storms. If the trunk is long the limbs will sway back and forth in the gale and not only break themselves but cause some of the under branches to break also by their dropping down upon them with their load of fruit.

The windfalls are much fewer and are more marketable where the trees are headed low. The fruit that drops, only has a short fall. The difference in the wind falls and their value alone should be enough to encourage the practice of low heads. It sometimes happens that windfalls from low headed trees may be disposed of at a fair price. While those that have dropped from the high heads on the hard ground are bursted or bruised so that they are unfit for market at any price, and must be hauled out or disposed of at an additional cost.

Many growers fail to realize the damage that may result from the ladder marks made in gathering the fruit. The higher the heads of the trees the more apt we are to injure their limbs by the use of heavy ladders.

Where the trees are headed low it will take much less cultivation than where the whole surface is exposed, as in the case of the high

headed trees. It is of great importance on many of our steep, shaly hillsides to shade the ground as soon as possible with the growing trees, as we can often retain the moisture much easier in this than any other way.

In many instances on very steep, thin land, as we said under "Self Mulched Trees," it has been found advisable to form a mulch around the trees by allowing their own limbs to cover the ground so closely that there would be no vegetation under them, thus solving the problem of mulching on these steep, thin lands, where hauling in a foreign mulch is not practical. Where land is so thin that all the plant growth gotten is needed on the space between the trees, this self-mulching has proven very satisfactory.

Having chosen the low headed tree as our ideal for steep lands, let us next consider some of the necessary requisites of this tree in order that it may be as complete an ideal as possible. Let us think of the fruit tree as a scaffold. One of the first requisites of a scaffold would be strength, so we are to build it low and strong. Then a two or three storied scaffold will hold more than a single story. As a fruit grower wants as large a yield as possible, we will build our trees, or scaffold, two or three stories high, thus largely increasing the capacity of the trees.

There is one thing we should always bear in mind, that the switches on our trees as they come from the nursery never produce fruit. Their function is to produce and hold up other wood that will in turn produce more switches. Finally we see the fruit buds forming on wood that is three, four or more years removed from the body of the tree. This being the case, we should think of these limbs that we find on the two-year-old tree as it comes from the nursery as the scaffold limbs, the foundation so to speak, upon which we are to form or build our tree. In order that this scaffold be strong, we should be careful not to leave these parts too long, remembering that the further from the body we get the load the easier the limbs will be broken. If you were to allow a switch three feet long to remain upon a young tree when set and it should make from its tip a growth of three feet, the first season, and the same the second, it would not take a very heavy weight at the end of this, then a nine-foot limb, to break it down. But if it had been cut back the first year to four inches, the second to six or eight, and to ten or twelve the third season, it would have been able to have held up a much heavier weight. Then in the building of our trees we should make the cuts short, also bear-

Typical Growth of York Imperial. Fifteen Years Old.

Typical Growth of Mammoth Black Twig
Fifteen Years Old

Typical Growth of Ben Davis
Fifteen Years Old

ing in mind that the tree will be stronger if each limb is growing on one a year older than itself. This is one thing against the practice of summer pinching, as it causes the multiplication of branches on a stem which grew the same year they did. So if we expect to build our trees for strength we should prune annually from time of setting, until the tree, or scaffold, is completed.

The manner or style of building will depend largely upon the habit of growth of the tree; for instance, the Transparent with its upright growth should have longer portions or parts left in its building than the Mammoth Black Twig, which is inclined to sprawl more like a Burbank plum. If the cuts are made short in the Transparent the top will be too thick and close, while if they are made long in the Black Twig its head will be even more open than is necessary, and the tree will cover too much ground and not have bearing wood throughout its center, as it would if held in check by cutting back.

PURPOSE OF PRUNING.—The first purpose of pruning should be for the general welfare of the plant. A good example of this would be where the gardener practices shearing off the tops of his celery plants two or three times

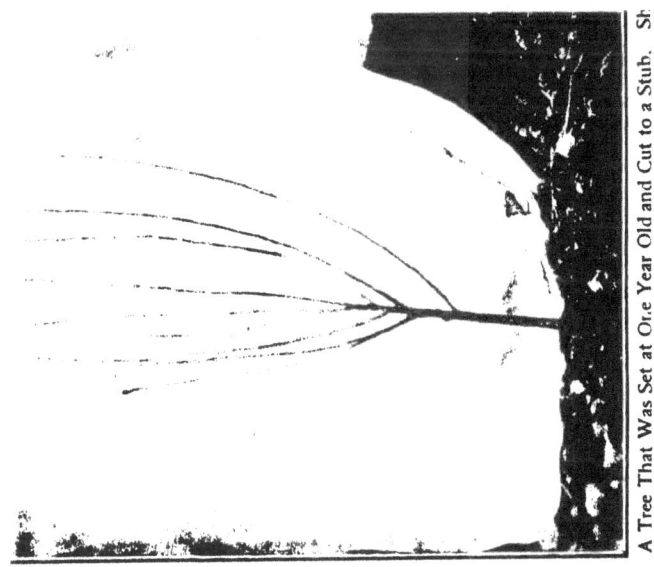

A Tree That Was Set at One Year Old and Cut to a Stub. Showing Growth of the Following Year Which is Cut Back to Form First Head With Center Left or Second Head to Form On

before they are transplanted, in order that the plants may grow more stocky and strengthen their root system. It is for the general welfare of the plant that the young tree is pruned when transplanted; for it often happens that the root system is much disturbed, in fact it is not uncommon for much of the nursery stock to lose one-half or in some cases even more, of its root system in being dug. Then it becomes necessary in order that the top and root may be balanced to use the pruning shears freely.

THE FIRST EFFECT.—The first effect of pruning is the multiplication of branches. This is particularly noticeable in young trees. For instance take the yearling switch as soon as its top is removed, let it be ever so small, a portion of the buds at once begin their growth and frequently we find branches starting along almost the entire length of the body. If we as orchardists would only bear in mind this first principle of pruning—the multiplication of branches—we make many cuts we never would have made, and neglect some that should have been made.

PRUNING TO MODIFY VIGOR.—We sometimes see and frequently hear of trees that are making too much growth. This often happens in the rank growing shade trees, and they become top heavy in spite of the lopping off of the branches every spring in an attempt to correct this heavy growth of wood, which will likely only be increased by severe pruning if done while the tree is dormant. Such trees should be pruned after the leaves are full grown. This will check their growth because it lessens their ability (as has been said, by decreasing their leaf surface) to prepare plant food which has been brought up from the soil to the leaves where it is made ready for the use of the tree. When we wish to modify the vigor of our fruit trees, in order to cause them to bear fruit, we should, as in the case just described, avoid winter or dormant pruning. That we may fully understand why we sometimes prune trees to cause them to bear fruit, we shall consider first the purpose of all plants. They grow in order to produce seed, or to reproduce. On observing plants we find that whenever their lives are threatened, that it is then an effort is made to produce seed. We should also remember that the apple tree does not start out to produce the apple, but its great purpose is to produce the seed, and it just so happens that its seed case—the apple—is made use of by us for food. So we find ourselves interested in the production of these cases. We have found under the subject just studied that it is possible to check the growth of trees by pruning them at certain times. We have also said that when the lives of trees were threatened they attempted to accomplish their mission, namely, produce seed. So we learn that we may cause a tree to produce fruit by threatening its life. This may be done in many ways. The bark may be split from the limbs to the ground, as many of us have seen practiced. The body may be driven full of nails, or we might prune their roots. Any or all of these and many other things might be done, which would threaten the life of the tree and cause it to attempt to reproduce, or to bear fruit.

CHECKING GROWTH TO CAUSE FRUITFULNESS.—When we threaten the life of a tree the wood growth is checked and the fruit buds are caused to form just as was described in the chapter on cultivation, under the head of "Growing Plants to Check Growth of Other Plants." So it is in regard to pruning to cause trees to bear fruit. The work must be done at a time to check the growth. As in the case of the shade tree, where we found its growth was checked because it was deprived of part of its leaves, and when we come to prune to cause it to bear fruit, we will pursue the same course by removing the limbs—that may be spared—after the leaves are full grown, thereby decreasing its ability to use plant food by lessening its leaf surface. By thus checking its growth or threatening its life, it is caused to attempt to reproduce. Summer pruning should only be resorted to in extreme cases. That is, when trees continue to make wood growth at the expense of fruit buds, after other means, such as has been described under the head of cultivation or "Growing Plants to Check the Growth of Other Plants." When it is found necessary to apply this method of pruning it may best be done by going over the trees while they are dormant and carefully deciding which branches should be removed. Then with a small paint brush on a long handle or pole, mark the limbs with white paint, so that after the leaves are grown you will have no difficulty in deciding which should come out. Otherwise it may make a good deal of work, and frequently limbs are removed that should not have been.

PRUNING TO PRODUCE LARGER AND BETTER FRUIT.—Pruning to increase the size and quality of fruit is probably more universally recognized and practiced among the grape and peach growers than any other class of horticulturists, although it has merit, and should be applied to the production of apples as well. Frequently we might thin our fruits by pruning, and thereby increase both their size and quality. This may be best accomplished, in the case of the apple, by thinning out some of the small branches on the outside of the tree, especially is this true in the case of the Rome Beauty. For when this tree attains age, it gets very thick on the outside. If we would do this outside thinning and leave more of the small twigs and fruit spurs through the center we could increase the quantity, as well as the quality of our crop. In the case of the peach this pruning should consist of the shortening or cutting in of the previous year's growth, the severity of this should be governed by the judgment of the grower and condition of the fruit buds, as described under "Pruning with Regard to the Location and Formation of Flower Buds."

In the case of the grape, "Pruning to Produce Larger and Better Fruit," would mean both the thinning out of the canes of the previous year's growth and the severe cutting back of those which are left. This will also be taken up under the head of "Pruning With Regard to the Location and Formation of Flower Buds."

PRUNING TO REMOVE UNNECESSARY PARTS.—Many of us only prune our trees with the idea in mind of simply removing those parts or branches that we may consider unnecessary. While this should be one purpose of pruning, it should not be uppermost in our minds, for we have already tried to show some reasons why pruning should be practiced that are of far more importance. If we all could have that ideal tree in our minds from the time of planting on through the orchard's growth, and prune accordingly, we might have much less of this unnecessary wood to remove. The most careful grower, however, will make mistakes and often leave too many limbs on the young tree. It is hard to realize how much thicker it will look after the limbs have gotten large, say six or eight inches in diameter. We should then decide as early as possible in the life of the tree just which branches are to be of real use, and remove as promptly as possible all others. Remembering that by so doing we avoid the wasting of energy or growth, as well as avoid making many large wounds that would be necessary in after years.

PRUNING TO REMOVE INJURED PARTS.—The removal of injured parts by pruning applies to the young trees when being prepared for setting, when all bruised or broken parts, either root or branch, should be removed. It also applies to the cutting off of diseased branches, such as blight, which frequently occurs in the pear and sometimes the apple. We also apply this principle of pruning when we cut out the cankered spots, or limbs in our apple trees. Again in the practice of cutting out the limbs of plum or cherry that are affected by Black Knot. We frequently have to remove injured parts after heavy crops of fruit or wind storms, such as broken or split limbs, which should be done promptly.

> "Is there a tree grown old or weak,
> On which the accustomed store you seek
> In vain—though of a kind well tried—
> Let it still, in its place abide.
> But go to work and shapely trim,
> Cut off each crooked, drooping limb.
> Perchance the ruthless pruning knife
> Will start anew its waning life."

PRUNING TO RENEW BEARING WOOD.—The renewal of bearing wood by means of pruning has long been recognized and practiced by the grape grower, and we have a system, the name of which signifies the practice, called Renewal. The peach grower possibly is the next in appreciation of the opportunity afforded by this system, and has often taken advantage of it by renewing his peach trees by severe cutting back (we call it de-horning) in seasons when the crop fails. In fact the peach grower must ever have in mind when pruning, the importance of renewing the bearing wood, as it is only on wood of the previous year's growth that the crop may be borne. So if at any time or for any reason the tree should fail to produce new wood there could not possibly be a crop of fruit the following season.

This is not so marked in the case of the apple as it may extend its growth only a very little at the extremities of its twigs and then form fruit buds. While in the case of the peach there must be at least a small switch growth each year in order to have fruit buds, as described under the head of "Pruning with Regard to the Location and Formation of Flower Buds."

PRUNING TO RENEW BEARING WOOD IN OLD TREES.—The renewal of bearing wood in old apple trees is much more of a problem than in the peach or young apple. Being a problem, however, does not make it any the less important. We only have to look about the neighborhood, and frequently on our own farms, to find striking examples of its necessity. The work of renewal should be carefully planned. The individual tree should be studied. After having decided upon the changes to be made in that old tree which has not had a limb removed for years, and has grown to look more like a living brushheap than a fruit tree, we shall have to exercise judgment or the work will be overdone. The changes to be made should be brought about gradually. Take out a limb on this side, and another in some other part of the tree, choosing, if possible, those which are interfering the most with the remainder of the tree. The next year the work may be followed up and other limbs removed. After removal of large limbs from an old apple tree, we find their place is often filled with water sprouts. How many ever stop to think why these sprouts make their appearance? It should be remembered that the trunk and branches have been shaded for years, and when we cut away these limbs and let the sun in on the body, the tree at once attempts to shade or protect itself by throwing out these growths. This same thing may occur when trees are heavily loaded and the branches bend down so that the sun shines directly on the bare portions of either body or branch. Then we should not forget that these water sprouts have a purpose, and will perform a work if allowed to. If these are all removed others will make their appearance, and it is well that they are persistent, for it is upon them we must depend for the renewal of the bearing wood in the centers of our old trees. Then when limbs are removed and these sprouts make their appearance, let us remember first that they are attempting to protect the tree, as well as to renew or rebuild the portions removed.

TREATMENT OF WATER SPROUTS.—When a tree is in this condition we should study it and see how many of these water sprouts are actually necessary in order that the tree may be protected. If we find a bunch of, say four or five shading a certain portion of the tree, we may remove a part of them, and then applying the first principles of pruning—the multiplication of branches—we should cut the remaining ones back, leaving six or eight inches (as the position on the tree may suggest) of the young growth.

These stubs will throw out branches which will shade as much or more surface, as all the original cluster. Besides multiplying the branches by cutting them back, we have caused them to take a firmer hold on the body of the tree and they will not be so easily broken off. The shortening of the stem has added to its strength or ability to hold the weight of crops in future years. The next season these sprouts that have grown one year and then been cut

Where Pruning Has Been Practiced for Mere Form and Size

back have thrown out many branches. We should thin them out and head in again. By this time we shall find the growth has taken on a more substantial appearance, and we may not be surprised to find many of the small twigs and spurs developing fruit buds. After a few years of careful, thoughtful pruning, the old thick topped, empty centered tree, may be changed into a more open top full of young, thrifty bearing wood.

PRUNING ONLY FOR FORM AND SIZE.—This is practiced in the forming of hedges or in the training of evergreens or other ornamental shrubs or trees, and seldom if ever applies to pruning of fruit trees. This method of pruning is really the continuation of that first principle, multiplication of branches. It is by this method we are able to get the solid shapely form of the cedar or other evergreens that we wish to grow into some certain shape or form. It is by the same means that we are able to make the hedge a smooth, solid mass of young shoots. The failure to observe this principle and put it into practice early in the growth of our hedges results with their being open next to the ground so that they are almost worthless as fences.

PRUNING TO FORM THE HEAD OF THE TREE.—When we begin to study pruning in relation to forming the heads of our trees we shall find many things to consider. The length of the trunk, or where the lowest branches are allowed to form, should first claim our attention. It is while the tree is young that this should be decided, as it is impossible to shorten the trunks of old trees, limbs might be shortened or headed in, but the length of body will remain the same. At the same time we should choose the form of head desired, whether it be vase-shaped—that is, only having one set of limbs, and they starting from close together—or shall we build a top on a central stem, with the limbs distributed over a considerable portion of its length. Or is our ideal still another form of head, one with two or three stories—or sets of limbs—some distance apart on the main stem or trunk, each set forming a complete circle in the top of the tree, while from the center of the last story stands the central stem or leader. Whatever form of top or head we decide upon we should have it in mind at all times while pruning. If possible we should be able to see how this branch will look a year from now, after the new growth is formed, if this or that cut should be made.

PRUNING TO REMOVE INSECT INFESTED PARTS.—It frequently happens that after the orchard is planted that the trees are found to be infested with scale. In order to treat these trees—which consist of spraying—successfully and economically, it is necessary to reduce the size of the top as much as can be done without injury to the tree. It sometimes becomes necessary to prune such trees severely. This pruning consists not only of the removal of the entire limbs wherever they can be spared throughout the top, but often the shortening in of all the branches, thus reducing the wood surface to be treated, thereby cheapening the operation. The shortening of the branches has lowered the top, thereby enabling the work to be done more efficiently, and unless spraying for scale is done very thoroughly the trees cannot be freed from the pest.

PRUNING TO BRING INTO MANAGEABLE SHAPE.—There are orchards that have been neglected and after many years the owner attempts to bring this sprawling, or that high top tree into manageable shape. This should be carefully considered by the grower before undertaken. It may mean heavy loss of wood, and the same result will follow as in the case of pruning to renew bearing wood. In fact the two are very similar, except it is sometimes the case that a tree may be in manageable shape, and yet not have the young bearing wood that may be desired. On the other hand a tree may have an abundance of bearing wood and yet be so high that it is impossible to manage it, and its crop to advantage. Again to bring into manageable shape may mean the shortening in of many of the branches. It may be these cuts will have to be made some distance from the trunk, say ten or more feet. When limbs are cut off at that distance from the main body care should be taken not to allow the formation of too many branches at the end. These clusters that are thus formed not only get very thick but sometimes become too heavy for the limb on which they are growing, and when loaded with fruit frequently cause the breaking of the scaffold limbs. To avoid this we should always make our shortening in cut at a branch.

When we undertake to remodel our trees, no matter for what purpose, we should remember that changing the heads, especially the lowering of them, by cutting out the tops carries with it more or less danger and should be considered most thoughtfully.

PRUNING WITH REGARD TO THE LOCATION AND FORMATION OF THE FLOWER BUDS.—We

A One Sided, Badly Shaped Tree
Courtesy F. H. Ballou

The One Sided Tree Corrected by Pruning
Courtesy F. H. Ballou

The One Sided Tree the Next Season After Being Pruned
Courtesy F. H. Ballou

seldom stop to consider the location of the flower buds when we go to prune our trees or vines. We give but little thought to the habits of growth and formation of buds on the different varieties of fruit trees. Our attention is seldom called to the difference in their location on the different kinds of trees. Unless we recognize some of these habits, peculiarities or differences as related to both the art and science of pruning, we cannot hope to be successful in the building of our trees. At the same time, pruning them in such a manner as to get the largest amount of high class fruit in color, size and flavor.

ART AND SCIENCE OF PRUNING AS RELATED TO FRUIT BUDS.—When we think of art and science let us remember that the art is the knowing HOW and the science is the knowing WHY trees should be pruned, and the effect. We are fast coming to realize the importance of the science of pruning. Having become convinced that it is not enough to simply know how a limb should be cut off, but that we must be able to give a reason for doing the work at a certain time in order that certain results may be brought about. In no way is this more apparent than in the study of pruning with regard to the location and formation of the fruit buds. This subject has been touched upon under the head of pruning to cause to bear fruit as well as under the head of care and cultivation. But when we consider the location of fruit buds we may be surprised how little we have really seen all these years while working among our trees. Let us look at some of our fruit trees and vines and see where these buds are found. On the apple we find the fruit bud always a terminal bud, maybe on wood that is one or more years old. They may be located on a twig, or as we say, on a fruit spur. This may be so short that we will hardly recognize it as a twig. We should remember that whenever a bud bursts and begins to grow, be it ever so little, we call it a spur or twig. Then if we expect our trees to bear fruit along their branches after they attain age, how careful we should be not to remove all those little leafy shoots which we find throughout our apple trees. It is upon these we should expect to find our first fruit buds, as these growths will be shaded by the outer and stronger growing branches (or their life threatened), therefore as their wood growth is checked the fruit bud is formed.

When looking for flower buds on the peach we find them located on the SIDES of switches of one year's growth. For this reason the peach grower cannot even HOPE to have a crop of fruit unless his trees have made some switch growth the previous season. When the habits of growth of the apple and peach are compared, we find that if it were possible to go over an apple tree and remove all the terminal buds, it could not bear any fruit that season. While with the peach, we might not only remove the terminal bud, but a portion of the wood growth could also be cut away. It would not only then be possible for the peach tree to produce a full crop, but the shortening of the previous year's growth (and probably the removal of a portion of the fruit buds in the operation) may serve as thinning, and thereby improve the size and quality of the fruits. This comparison helps us to realize the great difference in the habit of growth, or the location of the fruit buds on these two of our most common fruit trees. Seeing that such a difference as this exists, how foolish for us to go into the orchard and apply the same methods of pruning to both the apple and peach. The terminal buds we see on our quince trees in spring are not fruit buds. They must burst and make a growth, on the ends of which the flowers and afterwards the fruit appears. In this case the fruit is borne on the ENDS of shoots of the SAME season's growth. In the grape we find the buds, which are located on the sides of the previous year's growth, are not fruit buds but send out canes on the sides of which the grapes are borne. So it is seen that the grape is borne on the SIDE of the wood of the SAME season's growth. If we take these four common fruits we shall find that they give us four distinct types of growth, or location of the flower buds. Then when we prune we should take these into consideration; remembering the apple is always a terminal bud on wood one or more years old. The peach always on the sides of wood of the previous year's growth. The quince on the ends of growth of the same season. The grape on the sides of wood of the same season's growth.

WHEN TO PRUNE.—After having considered the purposes of pruning, which we must admit are many, and should be carefully studied before the work is undertaken, the question which next suggests itself is when should the work be done. There are several things which may determine this. Generally speaking, pruning should be done just as near the growing season as possible. The sooner after a wound is made that it begins to heal, the better for the patient, let it be a person or a tree.

There are, however, some exceptions, such as have been noted under the head of pruning to cause to bear fruit, to check wood growth, etc. There are other exceptions that should be considered which arise from habits of growth. Those which come under this head are the grape, maples, etc. For years it has been recognized as injurious to prune grapes at a season which will cause them to bleed, so we prune them before the growth starts, as we commonly express it, before the sap rises.

But strange to say we have been slow to realize the damage which may be caused by pruning trees which suffer in the same way. The maples, both the hard (or sugar) and the soft (or water), are often pruned at a season when if the trees were in a sugar camp there would be holes bored in the former to cause the sap to flow, which is used for making sugar or syrup. Yet many of us saw off great limbs from these trees at this season, and the sap runs down their bodies and causes the bark to die for many feet. In fact, often the whole side of the tree is ruined. Such trees should not be pruned until the leaves are grown, when bleeding will not occur, as it would have done earlier in the season.

When considering the pruning of fruit trees, such as apple, pear and peach, we prefer to have these cuts made as near the growing season as possible so they may begin to heal at once. When cuts, especially large ones, are made earlier in the season and exposed to the dry winter winds, the evaporation is very great. We often find these wounds cracking, just as the wood, which has been chopped ready for use and after lying in the sun and air for a few days (although the weather may be cold) we find cracks in all directions from the heart of the stick. Whenever this occurs the water finds entrance and causes trouble. A good dressing of some kind will overcome this to a great extent. The dressing should always be used on large limbs, regardless of the time the pruning is done.

How to Prune.—Having already decided that pruning should be done as near the growing season as possible, and having given our reasons for it, having made some exceptions, namely, the maples and any other trees that bleed, we come to consider how these limbs should be removed, how the cuts shall be made, and what tools shall be used. If we could, we should always do our pruning with the knife while building the young tree, making all the necessary changes in its growth while the wood was young and the tree growing rapidly, so that the wounds would heal quickly.

We do not always have or keep our ideal tree well enough in mind, however, and we find that we have left this or that limb that should have come out a year or two before. By this time they have grown too large to be removed with the knife, so the pruning shear is used. We find their work more satisfactory when a shear is used which cuts from both sides, as this avoids bruising the limb on the under side, or the barking of the body as sometimes occurs when a shear or clipper is used where the blade cuts down on an iron jaw or lip. In later years of the orchard's growth, we may find that we have made several mistakes which will have to be corrected, or from some other cause large limbs may have to be removed, so that it will be necessary to use the saw. The ax or hatchet should never be used as a pruning tool. We prefer a fine tooth, narrow bladed saw, in a strong, stiff steel frame, with handle or handles—as some are so constructed as to be easily shifted to different length handles—to suit the convenience of the operator. A good, strong keyhole saw, or something on that style has proven very satisfactory where it can be conveniently used.

WHERE TO CUT.—The question of where to cut when removing limbs from our trees is one that has not had the careful thought given it, that its importance deserves and even demands. We might well learn a few lessons from nature by the way in which she builds the great trunks of the trees of the forests. Few of us probably ever think when passing through the woods and noticing the knot holes in the trees how they came about. Many boys suppose them to have been made by the squirrels which seem to have such happy and comfortable homes "In the very heart of the oak," the door of which was the knot hole. Those who have been engaged in the clearing of our forests, and in the manufacture of lumber, recognize where there are knot holes in trees there is likely to be dead, or even rotten wood, and the tree or log may have lost much of its value. It is because of this decay that takes place—often to such an extent that the trunk becomes hollow—that the squirrel is enabled to enter the formerly sound bodies of our great trees. Knot holes are the result of self-pruning, or as we say, Nature's pruning. In order to realize that Nature prunes, we have but to walk through the woods on a winter morning when the ground is covered with the snow fall of the night before. We shall not go far before we notice dead twigs, branches and even limbs of considerable size which have dropped from the trees. These might have, and most likely would have been passed by unnoticed had it not been for the snow, which revealed the smallest twig. By this we see that nature prunes. We also see that she has purpose in her work, and the effects are evident. One great purpose is the thinning of branches, or even the destruction of the entire plant in many instances. The effect of nature's pruning may be seen on every hand. The tall tree that stands in some deep hollow with its long smooth trunk, without a limb for thirty or forty feet, is one of the most beautiful specimens of the finished product of the art of nature's pruning. Should you doubt that this tree has been pruned you have but to split its trunk or open it with the saw, and there at its very heart are to be found the knots made and left by the limbs that covered this stem or trunk years, or it may be, generations ago.

Those limbs were necessary for the development of the tree in the early years of its growth. They matured their buds, burst forth into leaf, and those leaves performed their functions just as truly as do the ones which wave so proudly now from its lofty boughs. But they are gone, pruned away. They outlived their usefulness and were sacrificed. The tool that was used was shade. The overshadowing growth of the more thrifty branches caused them to die and drop to the ground as those we saw upon the snow-covered surface of the forest. Should we study the life of such a branch we might learn a lesson. We notice a certain limb which for some reason has ceased to grow. Those around it push on even more rapidly; soon it has not only ceased to grow, but it is no longer clothed with leaves and it dies shaded to death. In a few months or years it has dropped to the ground and all there remains to mark its former place is a stub. The great body of the tree has not ceased its growth, but continues to build layer upon layer of woody fibre about its trunk. As these layers are built over the surface they find an obstruction—a stub, the remains of a limb which has been pruned away.

Now if this limb was small the stub is short, and these layers will soon cover it over and it will be as one of those heart knots which we

Result of Bad Pruning as Shown by Knot Hole in Which Squirrel Had Built Their Nests saw when the trunk was split or sawed open. But if it happened to be a large limb, then the stub would likely be much longer. Year after year the wood growth would pile up around it. During all this time the stub is decaying while the wood is being built up several inches about the now rotten stub, which finally drops out, leaving the knot hole. As time goes on, unless a very rapid growth of the tree closes this opening, the decay which began in the dead branch will be carried down into the body of the tree, resulting as it frequently does, in a hollow, which squirrels or bees are ready to make use of as homes.

When we prune our trees it will be well for us to consider where to make the cut, for if we leave a stub which cannot heal over, it must eventually rot off and leave a hole in the tree; it endangers the life of the apple just as surely as it would that of the forest tree. The stub cannot heal because there are no leaves beyond the cut surface to elaborate or prepare the plant food, and send it back to build tissues to cover the wound. Whenever we leave long stubs we should not be surprised that they do not heal, but simply die and drop off. Then we should not cut too close. If we do, the wound is made larger than is necessary and will require more time to heal over, than if it had been made at the proper place. We should do well to be governed by the little rolls of bark that we find at the base of the limb where it joins the body. If this is left unmolested, cutting just up to it, this roll continues to grow and covers over the wound in a much shorter time than the same sized wound could have been healed without this natural agency.

When heading in young trees we find it of great advantage to make the cut near a bud, for if it be made some distance from the bud the switch will die back to a bud and leave a stub, or as they look to us, a tombstone to mark our carelessness.

If we go on heedlessly with this work as many have in the past we may have to call in the tree surgeon and dentist, and have them take our shade and even fruit trees in hand. They, by boring and chiselling out dead or rotten wood and filling in with concrete, may save some of our beautiful and valuable trees. It would be better for us to be more careful in our work when pruning, and do it in such a way and at such times, that by the proper application of some coating material to the wound this dentistry could be largely if not altogether avoided.

CUTTING LARGE LIMBS.—It should be with the greatest care that we remove the large limbs from our trees. There may be great damage caused by cutting too close to the trunk and allowing them to split down. The wound thus made on the body will be much harder to heal than the one made by the removal of the limb proper. We have found the best method is to make two cuts, that is, leave a stub and remove it afterwards. We do this by first sawing up on the under side of the limb at some convenient distance from the body, and as soon as it begins to bind, remove the saw and place it on top of the limb a little to one side of the cut already made, and saw down until the limb falls. Then we have a stub left, which may be removed with perfect ease to ourself and safety to the tree.

TREATMENT OF WOUNDS.—After the pruning is done we should not fail to coat all wounds that are an inch, and over in diameter, with wax or paint. The latter is generally more convenient, and when used should not be too thin, remembering that it is being used as a means of protection against moisture, and to prevent the drying or cracking of the wood caused by the escape of moisture. In order to do this it should be thick enough to form a coating as near water proof as possible. Large wounds should not be expected to be completely protected by one coat of paint, but a second and sometimes a third is advisable.

Where the saw is used we frequently see orchardists attempting to smooth the wounds, generally using a draw knife. We are convinced this should never be done. The operator is apt to pare the outside edge more than the center of the wound, thus leaving the center the highest it will be much longer healing over than if left level. Again the rough surface left by the teeth of the saw will hold more paint, wax or whatever material may be used, than will a smooth surface, hence affording a more complete protection.

REMOVE THE BRUSH.—After the limb has been cut it should be removed from the orchard. This is of much more importance than is generally recognized, judging from the methods practiced. Frequently we find these prunings used to fill up washes in the orchard. This is dangerous, as it often affords breeding places for injurious insects, as well as a cover for rabbits and mice, which may work havoc with the growing trees.

SUMMARY.—To sum up the subject of Pruning we would say have an ideal. When thinking of an ideal we call to mind an exhibit at one of our great World's Fairs. If we as orchardists could have our ideal as plain in our mind as those who grew and trained those specimens must have had, and would follow it with the same persistency as the work exhibited showed they had, our success would be assured. It was an exhibit of apple, peach and pear trees trained against the wall, or on trellises. They grew as flat against these objects as if they had been grape vines, notwithstanding this artificial or forced habit of growth was one as foreign to their nature as could be imagined, it had all been brought about without leaving a single scar, or mark of knife, shear or saw. We love to think of those living, growing ideals, not simply because of their beauty, but because we were able to look beyond and back of them and see some one who had his ideal. So firmly and plainly was this fixed in his mind's eye that he could anticipate the growth of every bud and when one formed which promised or threatened to go in the wrong direction it was removed. If a branch were growing too long, the end was removed and the branches thereby multiplied. This is not practical in our orchard work, but we could do well to approach it to some extent by having our ideal always before us. From the time the tree comes from the nursery row until it is a bearing tree in the orchard try to anticipate the growth of their branches so that they may be corrected as early in life as possible.

We should have a reason for the removal of every limb. If we are not able to give a reason, then leave it on the tree. Make the cut at the right time, and in the right place, remembering stubs cannot heal, and are liable to cause damage to the tree. Treat the wounds as though they were on any other living thing, for this is the way we should think of a tree. Then remove and burn all prunings.

Begin the pruning for the *welfare* of the tree, continue and finish with this ONE object in view.

> "Your older trees need extra care,
> Lest slow decay their strength impair.
> Remove at once the fungous growth,
> All withered limbs, nor yet be loath
> To clean out all superfluous wood;
> Sunshine and air will do more good,
> Yet, if success is to be won,
> This work must not be overdone."

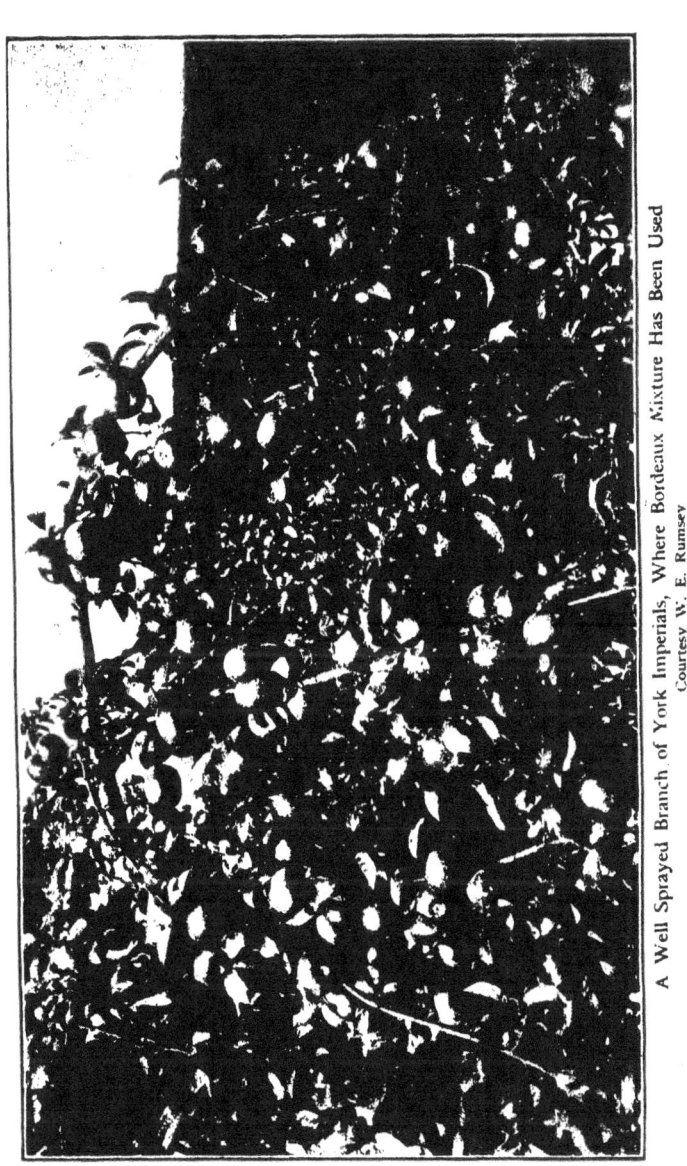

A Well Sprayed Branch of York Imperials, Where Bordeaux Mixture Has Been Used
Courtesy W. E. Rumsey

FRUIT FROM SPRAYED TREE

First Pile, Perfect   Second Pile, Curculio Stung   Third Pile, Wormy

FRUIT FROM UNSPRAYED TREE

First Pile, Perfect   Second Pile, Curculio Stung   Third Pile, Wormy
Courtesy W. E. Rumsey

CHAPTER XIII.

SPRAYING.

"If we have the scale from San Jose,
Or other pest, be what it may,
    Let us spray.
While we laze around they're feedin'
While we're swearin' they're a-breedin'
Plenty poison's what we're needin'
    Let us spray.

Lay a stock of Paris Green in,
Don't forget the kerosenin',
    Let us spray.
Into crevices go huntin',
Stop the hoppers' cheerful jumpin',
What we've got to do is pumpin',
    Let us spray.

> Pupa, chrysalis and millers,
> Fungi mixed with caterpillars,
>   Let us spray.
> Be they plump, or thin, or flaccid,
> Bring to bear on them the acid,
> That's the way to make 'em placid,
>   Let us spray."

The spraying of plants and trees has come to be recognized as not only one of the most important operations that the orchardists are called upon to perform, but an absolute necessity if we expect the business to yield either pleasure or profit. In many instances it has been found that without the use of sprays of some kind or other, it is impossible to preserve the beauty, or even the life of many of our shade trees, on our lawns and along the streets of our towns and cities.

It shall not be our purpose to say just when and what material we should use in spraying. This may all be gotten from the many splendid Bulletins which are being issued by the various Experiment Stations, as well as by the Department of Agriculture at Washington. Every fruit grower should keep in touch with these sources of information, remembering that these Bulletins are the results of the research work of many of the best men in the country. Men who are devoting their lives to the particular lines along which they write. We should deem it a great privilege to have these publications at our command. It shall be our purpose to give some suggestions to the practical grower as to why, when and with what to spray, and what appliances to use.

WHY AND HOW TO SPRAY.—Were we asked to give one reason for spraying we think the whole field, as broad as it is, might be covered by the one word PROTECTION. For we must protect the entire plant, stem, leaf, blossom and fruit. There are many things from which our plants should be protected. These troubles could well be divided into several classes, but for our practical use it may be enough that we divide them into but two great classes, Insect and Fungous troubles. Our remedies are also divided into two great classes, Insecticides and Fungicides.

If we are to give our plants protection against the many kinds of insects and protect them from the various fungous troubles our work must be timely. The old adage, "A stitch in time saves nine," never applied with more force than in the matter of spraying. The first thing for us to do then, should be to learn to recognize the troubles which are likely

Tanks for Mixing Bordeaux, Where Advantage Has Been Taken of the Hillsides By the Use of Gravity

to attack our plants, and know when to expect them to appear, that we may be able to meet or even head them off. We should think of spraying as a means of protection, or insurance, and as it is necessary to insure a building against fire before it begins to burn, just so is it necessary that the remedy be applied before our fruits become affected. After an apple begins to show Bitter Rot it cannot be saved by the application of a fungicide, but had it been covered with a fungicide—say a film of Bordeaux—then when the spore came in contact with its surface, it would have been killed at once and the fruit saved. In the case of the Codling worm, unless the poison is where the worm can get it when it attempts to enter the fruit it will not be effective, for when the worm is once on the inside of the fruit all the poison that might be applied could not hurt it. The first thing then to be considered in spraying is PROMPTNESS.

THOROUGHNESS.—After promptness the matter of thoroughness of application should be considered. As has been said, the work of spraying is for protection, and should be done before the trouble, whatever it may be, takes hold on the plants or fruits. If this work is to protect, it is absolutely necessary that all

Summer Spraying for Insects and Fungi
Courtesy A. L. Dacy

parts of the plants and fruits be covered. If one side of every apple on the tree should be kept covered with an effective fungicide, while the other side was left exposed, the entire crop might be lost from Bitter Rot or other fungous diseases. For if one side of the fruit is exposed the whole is in danger, so with all applications of sprays they must be thorough if we can even hope to be successful. When we think of why and how we spray we should remember that it means a thorough application of the proper materials applied in time to afford protection to our crops.

WHAT SHALL WE SPRAY WITH?—In order to know what to spray with we must not only know what troubles are likely to attack our plants and fruits, but we should have some knowledge of the ability of different plants to stand applications of certain materials of the various strengths. For example, the peach tree will not stand an application of Bordeaux the same strength as the apple.

Again we should take into consideration the time of the year as to the strength of the solution, whether the trees are growing or dormant, as well as the kind of material used. After considering the plant and its ability to stand the different mixtures at various strengths, we must next consider what material to use. Before we can do this we must call to mind what has been said about the troubles we are protecting against. We have said there are two great classes of troubles, fungi, and insects. For the fungous troubles we use some of the things which we find listed in our Bulletins and Spray Calendars as Fungicides; the chief of which has been for years the Bordeaux Mixture (Blue Stone and lime in various proportions), the 4-4-50 having been a standard strength.

We also have the Lime Sulphur Solutions, the home boiled, known as the California Wash; the chemically prepared Lime-Sulphur, and the Self-Boiled Lime Sulphur. These with the Bordeaux are the principal fungicides of the day and may be used as a protection against such fungous troubles as Apple Scab, whether it be found on leaf, flower, stem or fruit. We are persuaded that there has been more loss of crops caused by the scab in some of its form than any other fungi. Many orchards bloom profusely without setting any fruit. The cause is frequently attributed to cold winds, rains at blooming time, etc., while trees under similar conditions which have been sprayed have borne full crops. The real difference being that in one instance the flower stems—which are the future fruits, as the apple is the thickened stem—have been protected from the fungus causing the Apple Scab while the others were not, so when attacked dropped to the ground. The leaves on our trees must be protected from the same trouble so that they may be able to complete their work, thereby insuring vigorous buds the following spring. We all recognize the importance of protecting the fruit so that it will be free from that rough, brown spot that not only disfigures the apple, but checks its growth and causes it to become a one sided, knotty, worthless affair. We should just as surely protect the leaves, blossoms and stems of our trees.

We have other fungous troubles such as Rots. The chief among them is the Bitter Rot of the apple; Brown or Ripe Rot of the plum and peach. For the two last mentioned the self boiled lime-sulphur has proven the best remedy, not only for the rot in the peach, but to protect it against the Brown Speck fungus that causes so much loss to the peach growers. The self-boiled lime-sulphur is the recognized fungicide for the peach, while Bordeaux, or chemically prepared lime-sulphur is used for apple scab, for Bitter Rot we must still depend upon the Bordeaux mixture.

INSECTS.—Now we come to that other class of troubles, insects. There are a great many which attack our plants and fruits against which we must wage war. To list or classify these would be tedious, as well as useless for the average practical grower, so we shall divide them into two gerat classes, making this classification according to the way in which they take their food. First, the Sucking Insects,—those which take their food by means of puncturing the bark or outer portions of a plant and drawing on its sap,—such as San Jose and other scale insects; Aphis or plant lice which we find upon the leaves, stems and sometimes the roots of our trees.

SPRAYING FOR SCALE.—If oil is to be used to destroy the scale it should be applied while the trees are dormant. The weather conditions should be considered also; if possible the application of oil should be made on airy, sunny days, when the evaporation will be as great or rapid as possible, as this will very much lessen the danger of injury to the plants. We also have the Lime-Sulphur Solutions which are insecticides as well as fungicides, the home boiled known as the California Wash, and the chemically prepared lime-sulphur. These with the oil are the chief materials used in eradicating

Winter Spraying for Scale—A Cheap, Durable Outfit for the Farmer and Orchardist
Courtesy S. W. Fletcher

the scale. After growth has started scale can not be successfully treated with any of these materials as the necessary strength to be effective will damage the leaves and likely injure or kill our trees.

Secondly, we have the Chewing insects, such as the Codling Moth Worm, Potato Bug, Tent Caterpillar, Web Worm, and many others which devour plants or portions of plants.

> "Of bugs that chew, there's not a few,
>   Their poison must be eaten;
> But bugs that suck, outside we duck,
>   Or else they keep on eatin'.
> Then Paris Green, 'tis plainly seen,
>   Has got to go inside, sir;
> But kerosene with soap, to cream—
>   It kills them through the hide, sir;
> One ounce of soap, one quart of oil,
>   One pint of hot, soft water;
> One quarter hour churn up with power,
>   Dilute ten times you oughter;
> Of Paris Green one pound, I ween,
>   To each two hundred gallons;
> Then keep well stirred. It is inferred
>   These rules will save you millions."

It is of great importance that we know these insects. It is not necessary that we are able to call them by name, but we should know enough about them and their habits to be able to tell whether they belong to the sucking or the chewing class, in order that we may know how to protect against them.

A Cheap and Convenient Spray Outfit

Now that we have found such a vast difference in the way in which the insects take their food, we find that we must also use two kinds of insecticides. For instance in order to protect against chewing insects we cover the portions which are likely to be attacked with some poison, arsenate of lead being the most effective. But when attempting to protect against those which suck the sap from bark or leaf we find that poisons are not effective, so we have to use something which will kill by coming in contact with the insect; we call these, such as oil, soap, tobacco ooze, lime-sulphur, etc., contact insecticides. Air slaked lime and fine road dust are frequently used for the destruction of plant lice, as they kill them by stopping up the pores in their bodies through which they breathe, and these are also contact insecticides. We shall leave the exact time of making the various applications and the strength to use to be gotten by reference to the Bulletins and Spray Calendars to which we have already referred.

APPLIANCES.—After having decided that spraying is absolutely necessary in order to succeed and that it must be done promptly as a means of protection; also that the proper material must be used in a most thorough manner, there still remains a very important question to be considered, that of appliances, chief among which is the pump and its fixtures. We would not recommend any special pump, further than to say it should be a good one. The requisites of a good spray pump are many. The first we should mention would be ease of action; in order for a pump to be easily worked it must be a convenient height for the operator. We find that a pump on the side of a barrel is much more easily operated than when mounted on the head of a barrel, as this is too high for the convenience of the operator. Next after the position and ease of action of the pump we should see that it has a good agitator, one that moves with the slightest movement of the handle. Much of the efficiency of spraying depends upon the thorough agitation of the solution. We should select a pump with the ability to carry a heavy, steady pressure. Without pressure the work will not only be poorly done, but will necessitate the use of more material, thereby increasing the expense.

As has already been said, the secret of success in spraying is thoroughness of application. In order to accomplish this we should have a steady pressure. This cannot be obtained directly from the pump, but is the result of pressure stored up in the air chamber. So we should choose a pump with a large air chamber. Then if it should be necessary for the man at the handle to stop a minute to loosen a hose that has caught, perchance on a stub or rock, the power would not be noticeably weakened, but the operator with the spray rod would be able to do effective work for a short time at least even after the pump had stopped.

The pump should be supplied with a pressure gauge in order that we may be sure that we are getting a given amount of power. It is hard for the tired man to realize that the pressure is not as heavy in the evening, although he seems to be working just as hard as he was in the morning when he was fresh from a good night's rest. The gauge will tell the tale, and be of great value in aiding us in making the application uniform and thorough.

A pump should have sufficient capacity to admit of the use of two leads of hose, as it will economize time to spray two rows at once. We should see that the pump we choose is simple and easily taken apart, so that if anything should get out of fix it would not take too much time to relieve the trouble. All working parts should be brass so as not to be affected or injured by any of the solutions that may be necessary to use.

Long lengths of hose will be found to be very advantageous, especially on rough land, as they will make it possible to do the work with much less driving. Two or more rows may be reached at once, besides enabling the operator to reach up or down or possibly both ways on steep banks, and spray trees that could not be gotten at with shorter hose.

Extension rods are an absolute necessity in order to do easy and effective work. Either the brass lined bamboo may be used, or simply the quarter-inch gas pipe cut into the desired lengths. If the latter is used it will add very much to the comfort of the operator if a piece of old hose is slipped over the rod so that the iron will not come in contact with his hands. Besides making it warmer, it will also have a tendency to keep the liquid off of his hands.

In the selection of a nozzle we may be somewhat puzzled as there are many good ones on the market. The easiest and surest way to decide this is to settle upon what kind of spray we want. If a coarse one is desired, and it is to be applied under very heavy pressure, we might choose a certain kind. Or if a fine mist is wanted, and we expect to reach up to the parts and apply the spray from thte nozzle, we might choose another kind. If we wish to throw the spray in drops to considerable height something on the style of the Bordeaux should be chosen, which is particularly adapted to the spraying of extremely high trees.

All of these things should be thought over very carefully. Then the pump or appliances which come the nearest (according to our opinion) filling all these,—and it may be there are many other things that should be considered,—this is the outfit that should be purchased, taking into account the effectiveness and ease of the work as compared with the cost and durability of the pump.

THE STORY OF AN APPLE.

"A farmer picked this apple in his orchard in the west
And put it in a barrel with some others of his best,
Because they were so splendid, he declared the price must
　　climb,
And so he raised his figure on that barrel by a dime.

The man who bought that barrel stuck a label on the top,
Then told the interviewers of a shortage in the crop;
And when he came to sell it to a buyer on the floor,
He added on his profit and a half a dollar more.

The man who shipped that barrel stuck his label on it, too,
And talked of early freezes and the damage that they do;
The man to whom he shipped it said the grower's price was
　　high
And raised the price two dollars more than in the days
　　gone by.

The man who stored that barrel told of shortage in the pick,
Of scale and other pests that make the apple orchards sick.
And he put on five dollars to the cumulative price,
And so it went, each handler taking out his little slice.

So when you eat this apple, may it fill you with delight,
To know that some one profits on each nibble and each bite,
And O, be glad you do not live so very far away
From where the apple started, for think what you'd have to
    pay."

PACKING FRUIT IN THE ORCHARD
Courtesy F. E. Brooks

CHAPTER XIV.
PICKING, PACKING AND MARKETING.

"Lo, now we may gather of autumn's best treasure,
The fairest and sweetest and roundest of these;
With loved ones to help us, we heapen the measure,
And laugh in our joy as they fall from the trees—
The sweet juicy apples, the rich mellow apples,
The luscious ripe apples that fall from the trees."

After a careful study has been given all the various phases of the orchard business from the choice of sight, selection of the nursery stock, planting, care, cultivation, pruning and spraying, there is still room for complete failure unless the fruit is picked at the proper time and packed in a careful, honest and attractive manner. Many fruit growers have failed to make a financial success simply because they did not pay enough attention to this part of the

A Busy Scene, at a Busy Time of Year
Courtesy F. E. Brooks

business. They either became discouraged and quit the business, or financially disabled, so were compelled to seek other employment.

WHEN TO PICK.—The time of picking fruit should receive more careful consideration, as it affects not only the keeping of the fruit, but the quality as well. Many apples have been taken from the cellar in a shrivelled, wilted condition during the winter and spring simply because they were not allowed to remain on the tree until they were mature. These fruits were tough and tasteless because they had not been allowed to ripen, but were taken off when the grower happened to have the time and inclination to pick them. On the other hand much of our earlier ripening fruits are left upon the trees until they are over ripe, and reach the market or storage in bad condition.

There can be no set dates when certain varieties should be gathered. Soil and climate have much to do in determining the time an apple or other fruits should be picked. The Northern Spy, Baldwin and Greening are winter apples in New York State and should be gathered late in the fall, while in some sections of the Virginias these varieties must be gathered the last of August or first of September, or they will all drop. Thus we see that the climatic conditions govern or determine the time of harvest. When we consider some of the later maturing varieties that require a longer growing season in order to complete their growth, we find that the effects of climate and soil are just as marked in their case as in the other, although with the opposite results. Take a Ben Davis grown in a climate where the season is short—say New York—and you will get a tough, tasteless affair; while if grown on warm soil, where it will begin its maturity early in the season and the fruit is allowed to remain on the tree until late in the fall (and not gathered as soon as it has turned red) it finds a ready sale.

Again we shall find that the time of ripening of the variety may vary in the same orchard on account of the weather conditions. For instance, we may have to pick the Grimes Golden on a certain date this year while last year they were allowed to hang on the tree much later. The one year being very dry, therefore hastening the ripening of the fruit, while the fall before was seasonable and the fruits were not hastened in their maturity.

All these things have varying and decided effects upon the maturity of our fruits, and are of such importance that they should be considered and the fruit be allowed to remain on the tree as long as possible, and yet not long enough to become over ripe. There being no set rules to govern us, is it any wonder that we make so many mistakes in picking our apples?

BY WHAT SHALL WE BE GOVERNED IN PICKING.—In order that we may decide when to pick it is necessary that we not only take into consideration the climatic and soil conditions which have such marked influence upon the maturity of the fruit, but we should study the fruit itself, remembering that it is the seed case. As has been said before, the mission of all trees is to reproduce themselves. This reproduction is through seeds. It just so happens that we use the seed case (which we call fruits) for food. There is no better way for us to judge as to when the work of the tree is complete than when it has ripened its seed.

It would not be practical, however, neither is it necessary that we open the fruit and examine the seed every time in order to decide whether it is ready to pick. We may judge from the color of the case. In fact this coloring of the case seems to have a double purpose. First to indicate the seeds are ripe, as well as to make the cases attractive, so that they may be eaten and the seed set free, or that the cases may be carried to market and seeds distributed in other fields.

We see this seed carrying, travelling or distribution on every hand among the various plants. We should do well to pay closer attention to this coloring of the seed cases, realizing that they are saying by this change in their appearance that the seeds are ready to be set free and the case is now fit for food. We all know the effect of using them for food before they are ready. We only have to call to mind our experience with those green apples on the summer days when we were children and it all comes back clear and plain.

There is a great difference in picking fruits that are to be used at once, and picking for storage, where they are supposed to be held for considerable time. In fact when stored they are supposed to be used out of season. So we attempt in this case to arrest their development, or ripening, so as to prolong the life or keeping qualities of the seed case or fruit. Thus it is we find ourselves picking the Grimes Golden while they are still green, before the fruits have taken on their golden shade, just as they are clearing as we say. If they were allowed to remain on the tree until they were yellow or fully ripe, they would not keep when stored.

While this is the case with the Grimes Golden and other late fall apples that we may wish to hold for winter use, there are other varieties (especially of the longer keeping sorts) that should be left on the tree until the seeds are fully ripe and the cases have colored.

MAKE SEVERAL PICKINGS.—There are some varieties of apples which will pay especially well to pick several times. The Rome Beauty is one which we find pays to go over the trees as soon as there are some well colored specimens. If this is not done, many will drop and be lost. Again if the trees are very full we may be able to lighten their load and whenever the least weight is taken from their bending branches they change their position, thus allowing the sun to reach many fruits that have been partially, or quite covered before. Where this has been practiced we find that the quantity of no. twos have been very materially reduced. As this thinning process (for that is what it is) goes on, the remaining fruits grow very rapidly. It frequently happens that apples which have to be classed as no. twos when picked with the early ripening specimens would have made good no. ones, and sometimes fancy, simply by allowing them to remain on the tree until they had been given a chance to develop and ripen.

There are few of us who would go to our peach trees and gather all the fruit at a single picking. Yet we persist in taking the apples clean, whether they have grown on the outside limbs where the abundance of sun and light has hastened their maturity, or whether they have grown on an overloaded limb, and their position happened to be where the sun seldom shone, until after some of the fruit and much of the foliage had disappeared. A closer study and careful observation of this subject will do much to increase the value of our orchard products.

How to Pick.—Very much depends upon the way fruit is picked, much more than many growers seem to think. We often find pickers bringing baskets of apples to the tables almost covered with short twigs, which are growing to the apple stems. The packers frequently object to this because of the time and labor required to take them off, and if they are left on it makes a bad looking package. While these are both good reasons, there is still another of greater importance, one that the grower should consider. When we studied the Location and Formation of Fruit Buds in Relation to Pruning, we found that the fruit buds of the apple were on the end of the twigs or spurs. So when we see a bunch of apple blossoms they are on the end of a branch, but when we go to gather the apples they are very frequently at the base of a young shoot. This growth may not be very long. Its length will depend upon the vigor of the tree, strength of the land, cultivation, size of crop, etc. Then we should remember that the fruit bud of the apple is borne on the end of the branch. As the season advances, and while the fruit is growing this branch attempts to lengthen by pushing out a shoot from the place where the apple is borne. The mission of this twig will be to form a terminal bud which, if the vitality of the tree and other conditions are favorable, will become the fruit bud for the coming season. If the careless pickers in gathering the apples also break off this little shoot, they remove all possible chance for that twig to produce fruit the next year. It must spend next season in growing another twig and forming a terminal bud which may be a fruit bud, thus losing a season's work. We should be careful then when picking our apples not to pick two crops at once, and that is what has happened when the packing table is covered with those twigs.

We may avoid much of this trouble by holding to the limb with one hand and passing our finger or thumb over the fruit and dislodging it with a slight pressure on the stem, rather than simply to pull down on the fruit until it lets go and the limb flies up with a jerk, often causing some of the best fruit to fall to the ground bruised and practically worthless.

WHAT TO USE IN PICKING.—We prefer a smooth split basket with drop handle to pick in. Fruit seems to reach the table in better condition than when sacks are used. With very tender skinned varieties such as Transparent, Grimes and Delicious, it would be well to pad or line the basket.

Slight bruises in the orchard may mean a bad spot by the time the fruit reaches the market or consumer. We should be careful not to bark or bruise our trees when picking, as this may result in really more damage than spots made on the fruit. Care should be exercised in handling the ladders. They should not be leaned against the limbs in a slanting position, so that when the picker climbs up, the ladder will slip down and down, breaking twigs and bruising bark as it goes. These injured portions are inviting places for the spores to find winter quarters and may cause Canker spots on the limbs. Climbing into or standing in the forks of the trees frequently results in dead spots, and we are sometimes at a loss to know the cause. While taking care not to bruise or damage the fruit we should not for-

Preparing for the Harvest
Courtesy F. E. Brooks

get the tree upon which we must depend for future crops.

PACKAGES.—This is a greatly agitated subject. What package should be used, barrel or box? The barrel has been and will be the principal package for many years, especially for apples for the general market, while the box may be used for the fancy high quality varieties, which are to be used as stand apples in our cities and towns.

The fruit for the two packages differ very greatly. In the case of the barrel the fruit may vary in size and it will be of no particular disadvantage. In box packing it is almost impossible to make a neat pack without uniform sized fruit. This necessitates the thinning of our fruit in order to secure the uniformity in fruit necessary.

The question of labor must be considered in connection with box packing, as it will require extra help at this especially busy time of the year. For such fruits as sell best in boxes are tender, high classed varieties, which must be wrapped in order to reach the consumer in best condition. The boxing of the proper kinds of fruit will tend to increase the consumption of apples, as many families can use a box of fruit while it is in good condition, but would not attempt to use a barrel. A mistake is frequently made by putting apples of poor quality, and not up to the standard in size or color in boxes. This kind of box fruit is detrimental to the grower, and unsatisfactory to the consumer. The demand of our markets should largely govern us in our decision in regard to choice of packages. There are some markets that demand the half-bushel basket, especially with such early, tender skin varieties as Transparent, Dutchess, Early Harvest and Red June. Whatever is used it should be neat, attractive and honest throughout. The safest guide is to make each package so good that it will cause the consumer to ask for another from the same grower.

MARKETING.—After a crop is grown, picked and packed, we have the final problem to solve, and upon its solution the success of the entire business depends. That is the selling or the marketing of the product. It seems to us that there is a difference in selling and marketing a crop of apples. There are a great many growers who would make a good sale of their crops to the buyer or cold storage man who would make a complete failure if attempting to put the same apples on the town or city market day after day. We believe the average

On Our Way to Market
Courtesy W. E. Rumsey

grower would do better to sell than to attempt to market, or wholesale rather than retail.

TIME TO SELL.—Much will depend upon the season and upon the supply of fruit as to the best time to dispose of the crop. If a fair profit can be made by selling direct from the orchard we believe it is the best, allowing the storage man who is fixed to handle the fruit, and understands the market and its demands, to do the real marketing. Should we store and attempt to market our product in a retail way, we enter a field already occupied by men who have studied that branch of the business as thoroughly as we have the side of production.

WHEN TO MARKET.—There are many growers who do not stop to consider when an apple should be marketed. We find Ben Davis on the market in October and November, and the people decrying them. When if they had been kept and placed on the market in the spring, say March or April, they would have been in proper condition for use, and the people would have received them gladly. Among the many lessons that we as fruit growers should learn there are none that will have a greater effect on the consumption of fruit—if we expect to distribute or retail our crops—than for us as a class to learn when certain apples are at their best, and see that they are in the hands of the consumer at that time.

SPECKED APPLES.

"Old farmer Grump, with thrifty care,
  Had safely stored away
For winter use his apple crop—
  Enough to last till May.

'We'll not begin,' said Farmer Grump,
  'To eat 'em yet awhile;
They've got to last the winter through—
  There's none too big a pile.'

And so they lay 'neath lock and key,
  Till the ripest showed decay;
'Begin on 'em,' then the farmer said,
  'Begin on 'em right away.

'We'll kinder sort them out,' said he,
  'And use for sass the wust,
And everyone who goes for 'em
  Must take the specked ones fust.'

And so they used the specked ones first,
  As farmer Grump had said,
But though they ate some every day
  The specked ones kept ahead.

And they not only ate them first,
  But all the winter through,
If that's their way, I've naught to say,
  And naught I'm sure have you.

Now, Farmer Hearty also had
  A well-filled apple bin,
But as he stored them in he said,
  'Now, listen. We'll begin

To eat the best of 'em right off,
  And keep on so each day,
For some of 'em will not keep long,
  Though some will last till May.'

And so his household, one and all,
  Enjoyed the fruit while sound;
And eating still the ripest first,
  Had some when May came round."

THE APPLE BARREL.

"It stood in the cellar low and dim,
  Where the cobwebs swept and swayed,
Holding the store from bough and limb,
  At the feet of autumn laid.
And oft, when the days were short and drear,
  And the north wind shrieked and roared,
We children sought in the corner here;
  And drew on the toothsome hoard.
For thus through the long, long winter-time,
  It answered our every call
With wine of the summer's golden prime
  Sealed by the hand of fall.
The best there was of the earth and air,
  Of rain and sun and breeze,
Changed to a pippin sweet and rare,
  By the art of the faithful trees.
A wonderful barrel was this, had we
  Its message but rightly heard,
Filled with the tales of wind and bee,
  Of cricket and moth and bird;
Rife with the bliss of the fragrant June
  When skies were soft and blue;
Thronged with the dreams of a harvest moon
  O'er fields drenched deep with dew.
Oh! homely barrel, I'd fain essay,
  Your marvelous skill again;
Take me back to the past, I pray;
  As willingly now as then—
Back to the tender morns and eves,
  The noontides warm and still,
The fleecy clouds and the spangled leaves
  Of the orchard over the hill."

—*Edwin L. Sabin.*

Ladders Used in Picking High Headed Trees, as They Use to Grow Them

The Old Apple-Barrel
Courtesy Leo Jellinek

PLANT A TREE.

He who plants a tree
  Plants a hope,
Rootlets up through fibers blindly grope,
Leaves unfold into horizons free,
  So man's life must climb
  From the clods of time
  Unto heavens sublime,
Canst thou prophesy, thou little tree,
What the glory of thy boughs shall be?

He who plants a tree
  Plants a joy,
Plants a comfort that will never cloy.
Every day a fresh reality,
  Beautiful and strong,
  To whose shelter throng
  Creatures blithe with song.
If thou couldst but know, thou happy tree,
Of bliss that shall inhabit thee.

He who plants a tree
  He plants peace.
Under its green curtains jargons cease.
Leaf and zephyr murmur soothingly,
  Shadows soft with sleep
  Down tired eyelids creep,
  Balm of slumber deep.
Never hast thou dreamed, thou blessed tree,
Of the benediction thou shalt be.

He who plants a tree
  He plants youth,
Vigor won for centuries—in sooth,
Life of time that hints eternity.
  Boughs their strength uprear,
  New shoots every year
  On old growths appear.
Thou shalt teach the ages, sturdy tree,
Youth of soul is immortality.

286   *Practical Orcharding On Rough Lands.*

> He who plants a tree
> He plants love.
> Tents of coolness spreading out above
> Wayfarers he may not live to see.
> Gifts that grow are best,
> Hands that bless are blest,
> Plant—Life does the rest.
> Heaven and earth help him who plants a tree,
> And his work his own reward shall be.
> —*Lucy Larcom.*

A Prize Winning Plate of York Imperials

71—The Original Grimes Golden Apple Tree. This Famous West Virginia Seedling Originated on the Farm of Thos. Grimes, in Brooke County

TO AN OLD TREE.

"The suns of long summers have withered thy form
Around thee has beaten the pitiless storm;
The head that so proudly looked over the plain,
Is bowed with the burden of limb racking pain.

The friends of thy youth with a faint, dying moan,
Fell prone at Death's biding, and left thee alone,
And only a moldering grave mark of wood
Is seen where they once in their strong beauty stood.

"And soon or late to all that sow
 The time of harvest shall be given;
 The flowers shall bloom, the fruit shall grow,
 If not on earth at least in heaven."

THE ORCHARD LANDS OF LONG AGO.

"The orchard lands of long ago;
O drowsy winds, awake, and blow
The snowy blossoms back to me,
And all the buds that use to be,
Blow back along the grassy ways
Of truant feet, and lift the haze
Of happy summer from the trees
That trail their tresses in the seas
Of grain that floats and overflow
The orchard lands of long ago.

Blow back the melody that slips
In lazy laughter from the lips
That marvel much if any kiss
Is sweeter than the apple is.
Blow back the twitter of the birds—
The lisp, the titter, and the words
Of merriment that found the shine
Of summertime a glorious wine
That drenched the leaves that loved it so,
In orchard lands of long ago.

O memory, alight and sing
Where rosy-bellied pippins cling,
And golden russets glint and gleam,
As in the old Arabian dream,
The fruits of that enchanted tree
The glad Aladdin robbed for me.
And, drowsy winds, awake and fan
My blood as when it over-ran
A heart ripe as the apple grow
In the orchard lands of long ago."
—*James Whitcomb Riley.*

www.ingramcontent.com/pod-product-compliance
Lightning Source LLC
Chambersburg PA
CBHW050049230526
45470CB00004B/1459